SCIENCE
A CLOSER LOOK

Mc
Graw
Hill
Macmillan
McGraw-Hill

Program Authors

Dr. Jay K. Hackett
Professor Emeritus of Earth Sciences
University of Northern Colorado
Greeley, CO

Dr. Richard H. Moyer
Professor of Science Education and
    Natural Sciences
University of Michigan–Dearborn
Dearborn, MI

Dr. JoAnne Vasquez
Elementary Science Education Consultant
NSTA Past President
Member, National Science Board
    and NASA Education Board

Mulugheta Teferi, M.A.
Principal, Gateway Middle School
Center of Math, Science, and Technology
St. Louis Public Schools
St. Louis, MO

Dinah Zike, M.Ed.
Dinah Might Adventures LP
San Antonio, TX

Kathryn LeRoy, M.S.
Executive Director
Division of Mathematics and Science Education
Miami-Dade County Public Schools, FL
Miami, FL

Dr. Dorothy J. T. Terman
Science Curriculum Development Consultant
Former K–12 Science and Mathematics Coordinator
Irvine Unified School District, CA
Irvine, CA

Dr. Gerald F. Wheeler
Executive Director
National Science Teachers Association

Bank Street College of Education
New York, NY

Contributing Authors

Dr. Sally Ride
Sally Ride Science
San Diego, CA

Lucille Villegas Barrera, M.Ed.
Elementary Science Supervisor
Houston Independent School District
Houston, TX

American Museum of Natural History
New York, NY

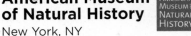

Contributing Writer

Ellen Grace
Albuquerque, NM

Students with print disabilities may be eligible to obtain an accessible, audio version of the pupil edition of this textbook. Please call Recording for the Blind & Dyslexic at 1-800-221-4792 for complete information.

The McGraw·Hill Companies

Macmillan McGraw-Hill

Published by Macmillan/McGraw-Hill, of McGraw-Hill Education, a division of The McGraw-Hill Companies, Inc.,
Two Penn Plaza, New York, New York 10121.

FOLDABLES is a Trademark of The McGraw-Hill Companies, Inc.

Printed in the United States of America

ISBN-13: 978-0-02-284135-5/2
ISBN-10: 0-02-284135-0/2

17 18 19 LWI 22 21 20

The American Museum of Natural History in New York City is one of the world's preeminent scientific, educational, and cultural institutions, with a global mission to explore and interpret human cultures and the natural world through scientific research, education, and exhibitions. Each year the Museum welcomes around four million visitors, including 500,000 schoolchildren in organized field trips. It provides professional development activities for thousands of teachers; hundreds of public programs that serve audiences ranging from preschoolers to seniors; and an array of learning and teaching resources for use in homes, schools, and community-based settings. Visit www.amnh.org for online resources.

Be a Scientist

Scientific Method

Observe

↓

Ask a Question

↓

Make a Prediction

↓

Make a Plan

↓

Follow the Plan

↓

Record the Results

↓

Try the Plan Again

↓

Draw a Conclusion

Life Science

UNIT A Plants and Animals

UNIT B Habitats

Earth Science

UNIT C Our Earth

UNIT D Weather and Sky

Physical Science

UNIT E Matter

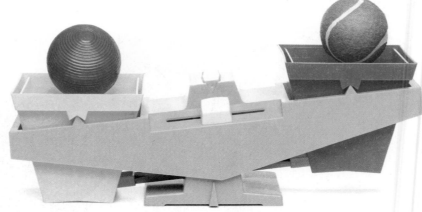

UNIT F Motion and Energy

Activities and Investigations

Life Science

Earth Science

Activities and Investigations

Physical Science

Be a Scientist

Some tree frogs lay their eggs
on leaves floating on water.

Look and Wonder

Do you see the frog? How does it stay on the lily pad?

How can a frog float on a lily pad?

What to Do

1 **Predict.** Where should you place the frog on the lily pad so that the frog stays dry?

2 **Make a Model.** Color a paper plate green with crayon. This will be the lily pad.

3 ⚠ **Be Careful.** Poke a small hole near the edge of the lily pad. Tie a six-inch piece of string through the hole.

4 Place the lily pad in a pan of water with the string below it.

5 **Record Data.** Draw a picture to show where you placed the frog.

You need

paper plate

green crayon

scissors

string

pan of water

toy frog

Step **4**

What do scientists do?

Scientists use many skills when they work. You wondered about the frog on a lily pad. Just as you did, a scientist might **make a model**. A model shows how something in real life looks.

Scientists use other skills that you can use, too. Scientists **observe**, or look carefully. A scientist who observes a pond can find many amazing things.

Scientists observe the height, color, and shape of plants near the pond.

cattails

pond grass

water iris

Scientists **compare** things by telling how they are alike or different. Look at the two pond animals on this page. How might a scientist compare them?

Look closely. Both animals have wings. They both live near ponds. But they are different in many other ways. Scientists find ways to **classify** things, or put them in groups. Insects and birds are different animal groups.

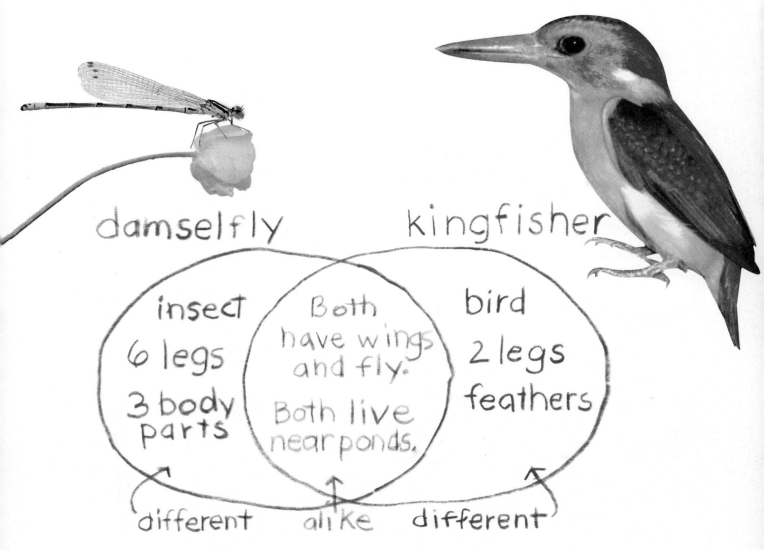

damselfly

kingfisher

insect
6 legs
3 body parts

Both have wings and fly.

Both live near ponds.

bird
2 legs
feathers

different   alike   different

How do scientists work?

Look at all the eggs a scientist found near a pond! Scientists can **measure** how large or how heavy the eggs are. When you measure, you find out how long or how heavy something is. You can also find out how hot or how cold something is.

The facts scientists find are called data. When scientists **record data**, they write down what they observe.

How long?

turtle	3 centimeters
frog	3 millimeters
duck	5 centimeters
robin	2 centimeters

After scientists collect data, they can put their data in order. **Put things in order** means to arrange them in some way. For example, you can order the eggs by their size. Which egg is smallest? Which is largest?

Another skill scientists use is **infer**. When you infer, you use what you know to figure something out. Can you infer which eggs belong to the animals on this page?

frog

duck

turtle

robin

How do scientists learn new things?

Scientists learn new things by investigating. When you **investigate**, you make a plan and try it out.

Scientists start by asking a question. They predict what the answer might be. When you **predict**, you use what you know to tell what you think will happen.

Look at the pictures of the tadpole and young frog. What do you predict the young frog will look like next?

tadpole

young frog

?

?

When you draw conclusions, you use what you observe to explain what happens. Scientists draw conclusions. They conclude tadpoles live in the water, grow legs, and climb onto land.

Scientists communicate their ideas to other people. When you communicate, you write, draw, or tell your ideas.

September 17
My Frog Notes

← head
← tail

First, it was a tadpole.

legs

Then the tadpole grew legs. It still has a tail.

short legs

no tail

long legs

Now it has long back legs and no tail.

My Conclusion:

Frogs grow legs and can walk on land.

Think, Talk, and Write

1. Which skill helps scientists put things into groups?

2. Write about what new things you might want to learn if you were a scientist.

Look and Wonder

This frog can swim! How else can a frog move? Scientists ask questions like this. They follow certain steps to find the answers.

How does a frog move?

What to Do

1 **Observe.** Look at the pictures on this page. Think about how the frogs are moving.

2 **Record Data.** Make a list of the different ways you see the frogs moving.

3 **Draw Conclusions.** Add to your list. Write the body part the frogs use to move in each way.

4 **Communicate.** How do frogs move?

How high can a frog jump?

Scientists investigate by following steps called the **Scientific Method**. Here is how one student scientist follows the Scientific Method.

Observe
Lola uses her science skills to observe the frogs in her classroom.

Ask a Question
Lola's question is:

Does a frog's size affect how far it jumps?

Make a Prediction
Lola predicts the answer is yes. She thinks Andy will jump farther because his legs are longer.

Andy

Molly

Make a Plan

Lola writes down a plan to test her idea.
When she writes the plan, other people
can follow it too.

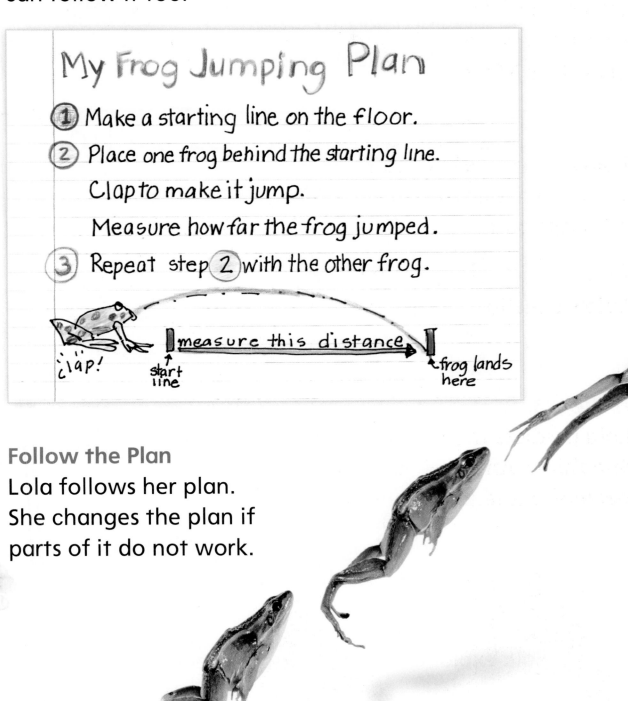

My Frog Jumping Plan

① Make a starting line on the floor.

② Place one frog behind the starting line.

   Clap to make it jump.

   Measure how far the frog jumped.

③ Repeat step ② with the other frog.

clap!   start line   measure this distance   frog lands here

Follow the Plan

Lola follows her plan.
She changes the plan if
parts of it do not work.

What did you find out?

Record the Results
Lola makes a chart to show
how far each frog jumps.

How far can each frog jump?			
frog	1st try	2nd try	3rd try
Andy	20 cm		
Molly	25 cm		

Try the Plan Again
Lola tests each frog three times. This helps
her know if her results are correct.

Draw a Conclusion
Lola explains what her results mean.

Lola talks to her classmates about what her results mean. This can lead to new questions and new investigations.

You can follow the Scientific Method when you investigate too!

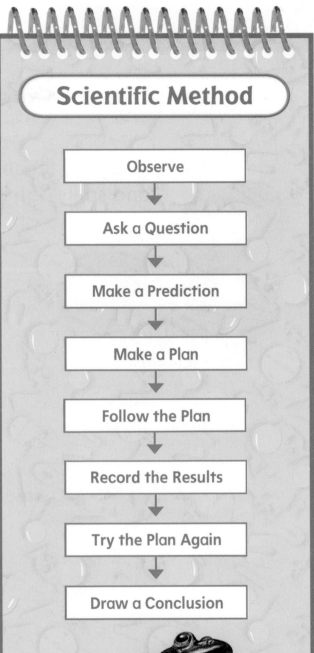

Scientific Method

Observe

↓

Ask a Question

↓

Make a Prediction

↓

Make a Plan

↓

Follow the Plan

↓

Record the Results

↓

Try the Plan Again

↓

Draw a Conclusion

Think, Talk, and Write

1. Why do you think it is important for scientists to make a plan?

2. Write about why scientists write down their plans.

Safety Tips

When you see △ **Be Careful, follow the safety rules.**

Tell your teacher about accidents
and spills right away.

Be careful with sharp
objects and glass.

Wear goggles when
you are told to.

Wash your hands
after each activity.

Keep your workplace neat.
Clean up when you are done.

Plants and Animals

Woodchucks eat plants and sharpen their claws on trees.

The Seed

A seed is so small,
it is hard
to believe
That it could
grow tall,
like the biggest oak trees
Or become a red rose,
growing on a long vine
How does the seed know,
that it will turn out
so fine?

Talk About It

What do you know about seeds?

19

Plants

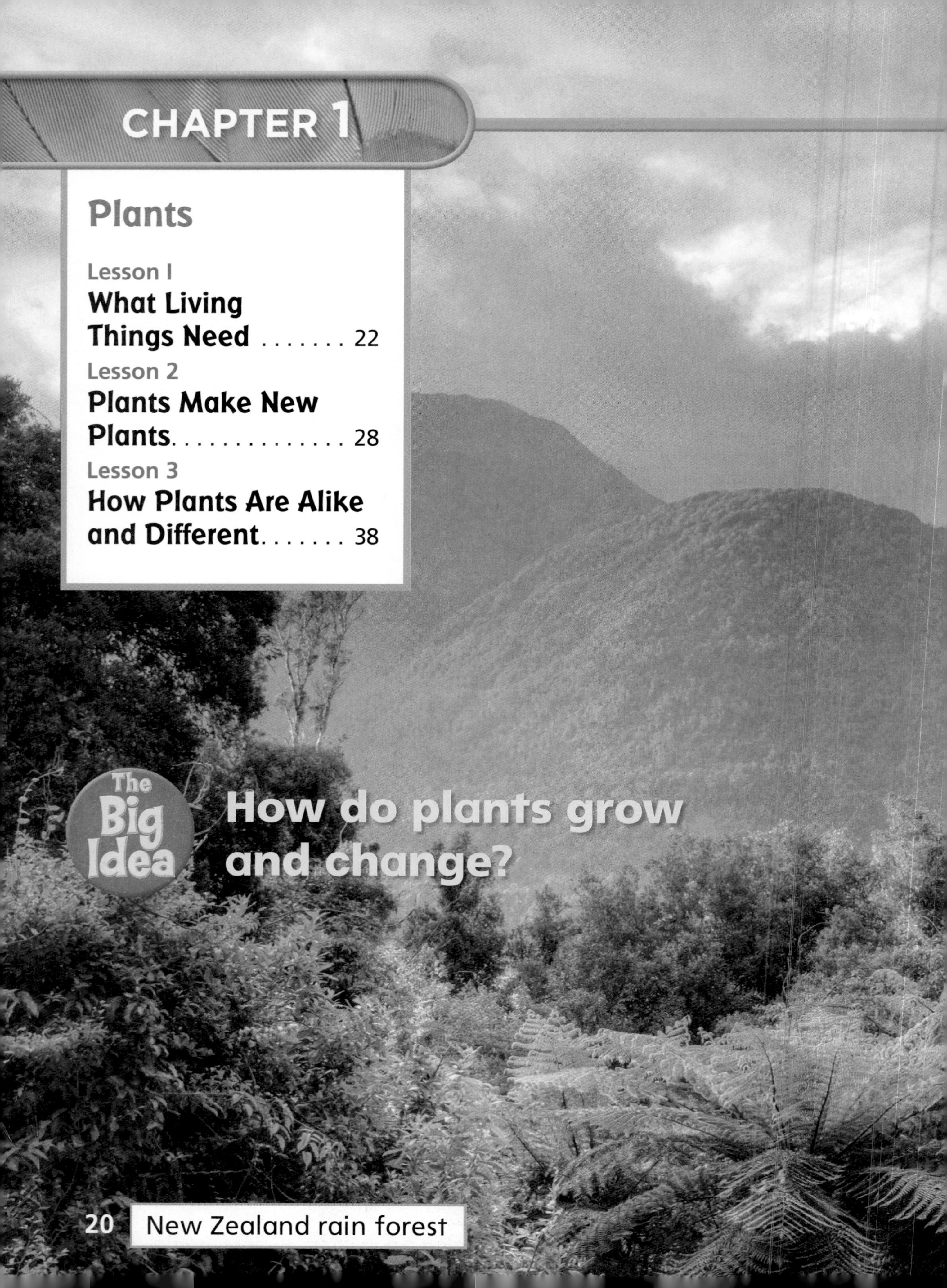

The Big Idea How do plants grow and change?

Key Vocabulary

flower plant part that makes seeds or fruit (page 30)

seed plant part that can grow into a new plant (page 30)

pollen sticky powder inside a flower that helps make seeds (page 30)

seedling a young plant (page 34)

21

What Living Things Need

Look and Wonder

What things in this picture are alive? How can you tell?

What do leaves need?

What to Do

① Put the plants in a sunny place. Choose one plant and cover its leaves with foil. Keep the soil moist in both pots.

② **Predict.** What will happen to each plant in a week?

③ **Record Data.** Write down what you observe for a week.

④ Were your predictions correct? What do leaves need?

Explore More

⑤ **Predict.** What will happen if the foil is removed? Observe the plant for a week. Was your prediction correct?

You need

two potted plants

foil

Step ①

What do living things need?

Living things grow and change. Sometimes it is easy to tell when something is living. You can see animals move, breathe air, eat food, and drink water. It might be harder to tell, but plants are living things, too.

▶ **A grasshopper eats a dandelion flower.**

The swan makes a nest for her chicks near a pond.

You have to watch plants over time to see them change and grow. Like all living things, plants need air, water, and space to live and grow. They also need food. Plants make their own food.

 What makes living things grow?

▽ **This sunflower takes most of the summer to grow into an adult plant.**

sprout

young plant

adult plant

How do plants make food?

Plants have parts that they use to help them make food. Plants need sunlight, air, and water to make their own food. Plants also need minerals. **Minerals** are bits of rock and soil that help plants and animals grow.

≡**Quick Lab**

Observe a plant. See what parts take in water.

Plants Make Food

Leaves take in air and use sunlight to make food. ⎯⎯

The stem holds up the plant. It allows water and food to travel through the plant. ⎯⎯⎯

Roots hold the plant in the soil. They also take in water and minerals. Roots can store food for the plant, too.

Read a Diagram

How do the parts of the plant help it get what it needs to make food?

When plants make food they give off a gas called oxygen into the air. **Oxygen** is what humans and other animals breathe in order to live.

These plants make oxygen that the boy and the dog need to live.

 What do plants need to make food?

Think, Talk, and Write

1. **Compare and Contrast.** How are plants and animals alike? How are they different?

2. What do roots, stems, and leaves do?

3. Write about how you can tell that a plant is living.

Art Link

Draw how a seed grows. What direction do the roots grow? What direction do the stem and leaves grow?

LOG ON **e-Review** Summaries and quizzes on line at **www.macmillanmh.com**.

Plants Make New Plants

Look and Wonder

Where do you think the seeds in this plant are?

What are the parts of a seed?

What to Do

① **Observe.** What does the outside of a dry lima bean feel like? Use a hand lens. What do you see?

② **Predict.** Draw what you think is inside the seed.

③ Use your fingernail to open the wet seed. Use your hand lens to observe the wet seed. Draw what you see.

④ **Communicate.** Compare your two drawings. What was different? What was the same?

Explore More

⑤ **Observe.** Look at other wet and dry seeds to see how they compare.

You need

dry lima bean

wet lima bean

hand lens

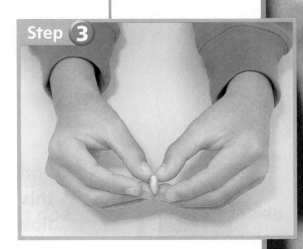

Step **3**

How do seeds look?

Most plants have seeds to make new plants. Seeds have food inside them to help the new plant grow. There are many different shapes and sizes of seeds.

Some seeds are small. Wind or water can carry them away. Other seeds stick to the fur of animals and get a ride to a new place.

Quick Lab

Observe the seeds inside an apple. Talk about how the fruit protects the seeds.

◄ There are many anise seeds inside this star-shaped pod. The shapes of the pod and the flower are alike.

▲ A marigold seed is small and thin. It does not have much food inside.

Seeds have many parts. All seeds have seed coats which protect the seed. Seed coats also help keep the seeds from drying out. Some seeds also have hard shells.

 Why do you think some seeds have shells?

▲ Peanuts are seeds. They come from peanut plants.

The shell of a peanut is hard and light brown.

The seed coat is thin and dark brown.

This part is a tiny plant. It will grow bigger.

These parts give food to the tiny plant so it can grow.

FACT Seeds are living things.

How do seeds grow?

A **life cycle** shows how a living thing grows, lives, makes more of its own kind, and dies. The plant life cycle begins with a seed. Seeds need a warm place, light, water, and food in order to grow.

Life Cycle of a Pine Tree

Adult pine trees make seeds in cones instead of flowers.

The pine cones fall to the ground. Some seeds get moved to other places.

A seed sprouts and becomes a **seedling**, or young plant.

The seedling grows into an adult pine tree. It grows cones so it can make new plants.

Read a Diagram

What does a pine tree have instead of flowers?

LOG ON *Science in Motion* Watch a plant grow at **www.macmillanmh.com**

Most plants follow the same life cycles as their parent plants. Different kinds of plants have different life cycles. Some plants live for just a few weeks. Other plants live for many years.

 What will a pine seed grow into?

◀ **These flowers go through their whole life cycle in just a few months.**

▲ **Redwood trees take more than two years just to make cones.**

Think, Talk, and Write

1. **Sequence.** How do flowers make new plants?

2. How would you take care of seeds to help them grow?

3. Write or draw pictures to show the steps in the life cycle of a plant.

Health Link

We eat the fruit and seeds of many plants. How many can you think of? What other plant parts do we eat?

 e-Review Summaries and quizzes online at www.macmillanmh.com

Main Idea and Details

Read about a plant that uses wind to move its seeds. The main idea is circled. The details are underlined.

Dandelions

Dandelions use the wind to move their seeds. Dandelion petals dry out when the flower dies. Then the seeds are ready to come off the flower. The seeds have long light tufts that can float in the air. Wind blows the seeds. They land in places where new plants can grow.

✎ Write About It

Write a paragraph about a flower that you observed. Make sure you have a main idea and details.

Remember

The main idea tells what a paragraph is about. Details tell more about the main idea.

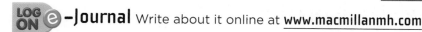

LOG ON e-**Journal** Write about it online at **www.macmillanmh.com**

How Many Seeds?

Some fruits, like watermelons, have many seeds. Other fruits, like peaches, have just one seed.

Solve a Problem

Suppose each apple on this tree had about 5 seeds. If you picked 3 apples, about how many seeds would you have? Show how you found the answer.

Write a number sentence about fruit seeds. Show your work.

Remember
You can draw pictures to help you find the answer.

How Plants Are Alike and Different

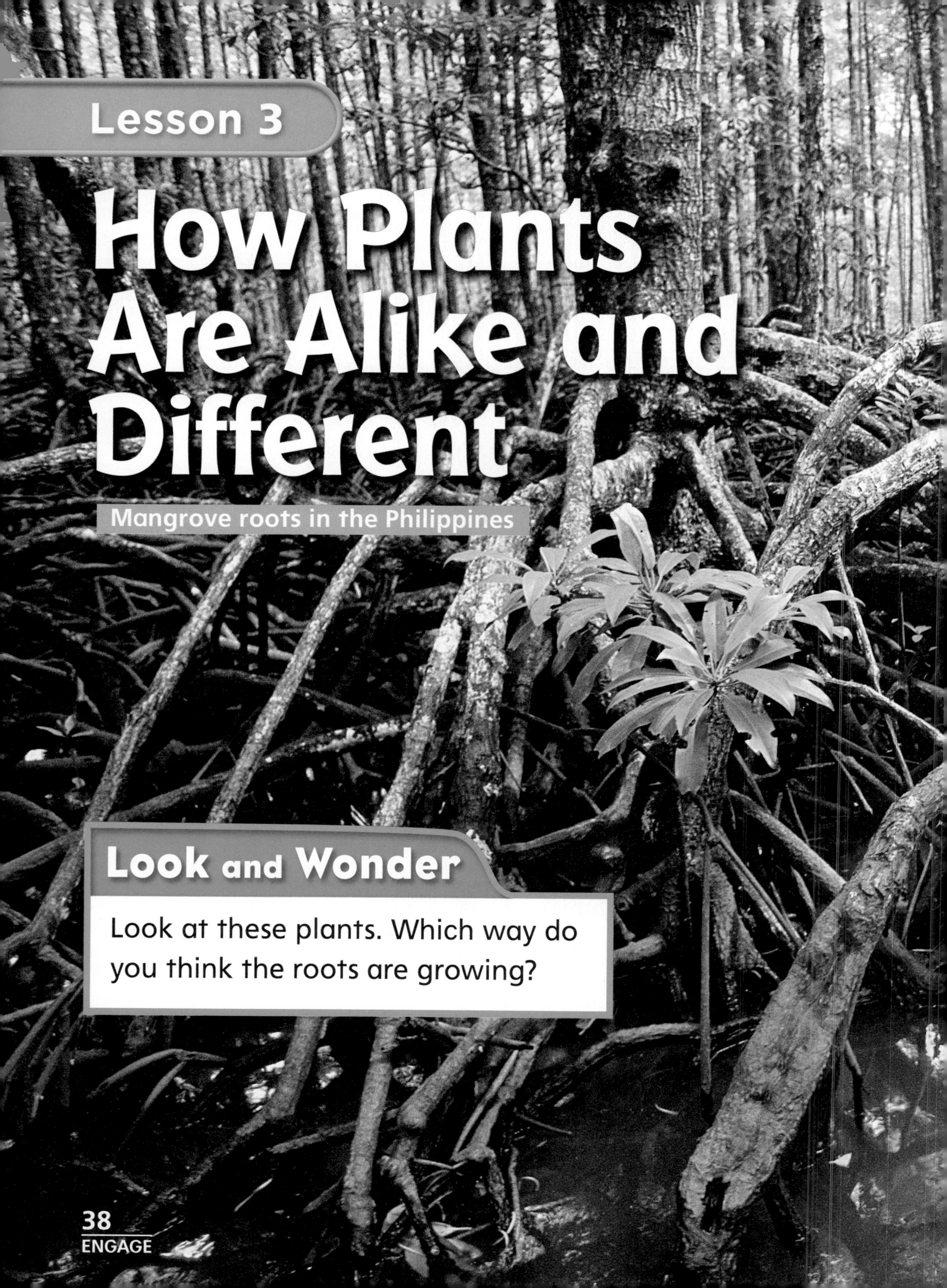

Mangrove roots in the Philippines

Look and Wonder

Look at these plants. Which way do you think the roots are growing?

How do roots grow?

You need

What to Do

① Put three bean seeds on a damp paper towel. Put them in the bag and tape it to a bulletin board.

3 bean seeds

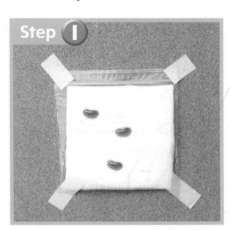

Step ①

paper towels

② **Observe.** Which part grows first? Which way did the roots grow?

tape

③ After the roots grow, turn the bag upside down. Tape it to the board again. Make sure the paper towel stays wet.

④ **Draw Conclusions.** What happened to the roots?

plastic bag

Explore More

⑤ **Investigate.** What happens to the roots if left in the dark?

hand lens

Vocabulary

trait

How are plants like their parents?

You know that cats have kittens and dogs have puppies. Animals have babies that look and act like their parents. Plants do the same thing.

▼ An acorn can grow into an oak tree.

▶ A sunflower seed can grow into a sunflower.

The way plants or animals look or act is called a **trait**. Young plants will have many of the same traits as their parents. Some plants might look a little different from their parents. The plant will still have the same shape of flowers, petals, and leaves.

▲ **Long ears are a trait of Basset hounds.**

✓ **What are some traits of a sunflower?**

Tulips

Read a Photo

How are these tulips alike and different?

How do plants survive in different places?

Plants change to get what they need from the place where they live. When a seed begins to grow, the roots always grow down. Plant parts may look different in different places, but their parts still help make food.

▲ This Joshua tree and other plants in very dry places have few or no leaves. These plants store water in thick stems.

▲ This banana tree and other plants in very wet places have large leaves. They help get light in the thick, dark forest.

Plants can change to stay safe, too. Some plants have ways to stay safe from animals. Other plants need to stay safe from the weather where they live. When plants change during their lives, those traits are not passed down to their offspring.

✅ **Why do you think some plants have thorns?**

◀ **On the coast, the wind is so strong that the branches on the trees bend.**

Think, Talk, and Write

1. **Classify.** Think of four ways that plants are like their parent plants.

2. What changes the way plants grow?

3. Write about the way a plant grows from a seed. How do the roots grow? Why?

Art Link

Make a crayon rubbing of two different leaves. How are they alike? How are they different?

The Power of Periwinkle

People who live in forests all over the world know about helpful plants. They use plants for food and for building homes. They also use plants to make medicine.

One helpful plant is the rosy periwinkle. It first grew in Madagascar, and later people spread it around the world. People now use the plant to treat fevers, sore throats, toothaches, and upset stomachs.

Today some forests in Madagascar are being cut down. People clear the land to grow food. Scientists want to keep these forests safe. There may be more helpful plants to study and use.

Madagascar

AMERICAN MUSEUM Ö NATURAL HISTORY

▲ Scientists and local people use the rosy periwinkle to treat diseases.

This woman gathers rosy periwinkle plants. ▶

Talk About It

Classify. Make a list of plants you know. Classify them by how they help people.

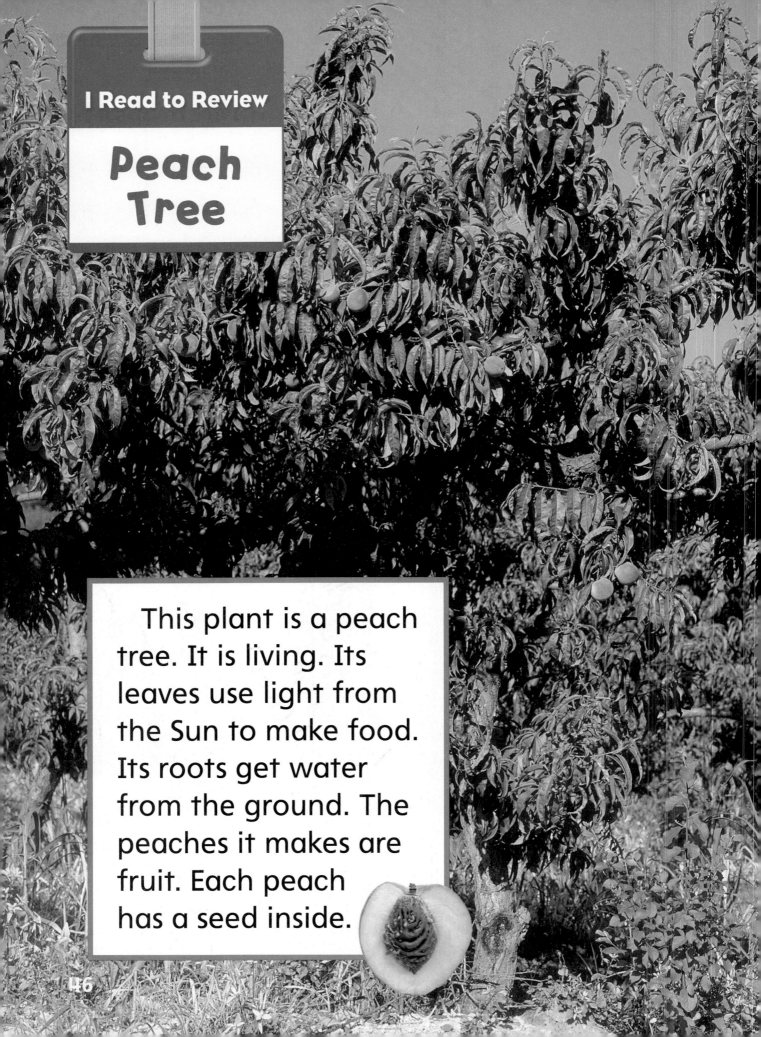

Peach Tree

This plant is a peach tree. It is living. Its leaves use light from the Sun to make food. Its roots get water from the ground. The peaches it makes are fruit. Each peach has a seed inside.

Seeds need water and a warm place to grow. The roots grow down into the ground. The stems and leaves grow up toward the light. This peach tree will grow to look like its parent plant.

The tree grows bigger and makes flowers. Flowers help the plant make new plants. Many of these flowers will grow into peaches.

This tree is covered in peaches. People and animals eat the peaches and leave the seeds behind. The seeds can grow into new peach trees.

CHAPTER 1 Review

Vocabulary

Use each word once for items 1–5.

flower

life cycle

pollen

seed

traits

1. A _____ shows how something grows, lives, and dies.

2. The part of the plant that makes the seed is called the _____.

3. The ways that plants and animals look like their parents are called _____.

4. Flowers need a sticky powder called _____ to make seeds.

5. This is a _____. It will grow into a new plant.

Answer the questions below.

6. Compare and Contrast. Look at the pictures below. What traits do these plants share?

7. What do seeds and seedlings need to live and grow?

8. Observe. Look at the plants in the picture below. Describe them.

The Big Idea

9. How do plants grow and change?

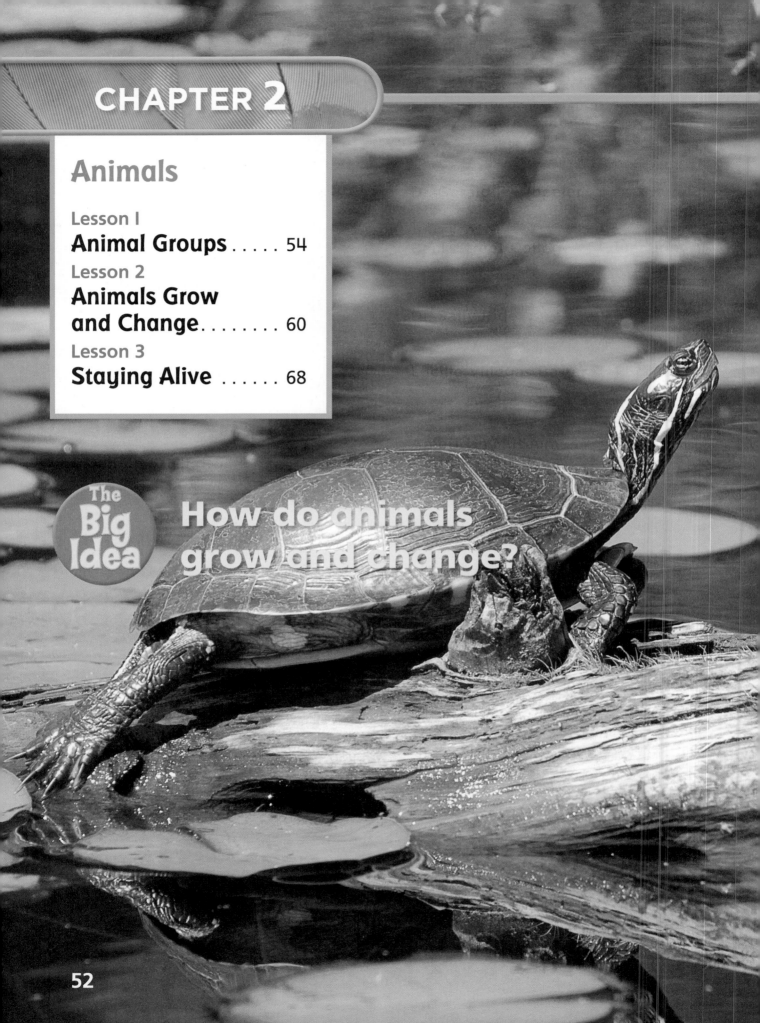

CHAPTER 2

Animals

The Big Idea

How do animals grow and change?

52

Key Vocabulary

mammal animal with hair or fur that feeds milk to its young (page 56)

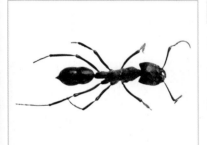

insect animal with six legs, antennae, and a hard outer shell (page 58)

larva stage in the life cycle of some animals after they hatch from an egg (page 64)

adaptation body part or way animal acts that helps it stay alive (page 70)

Animal Groups

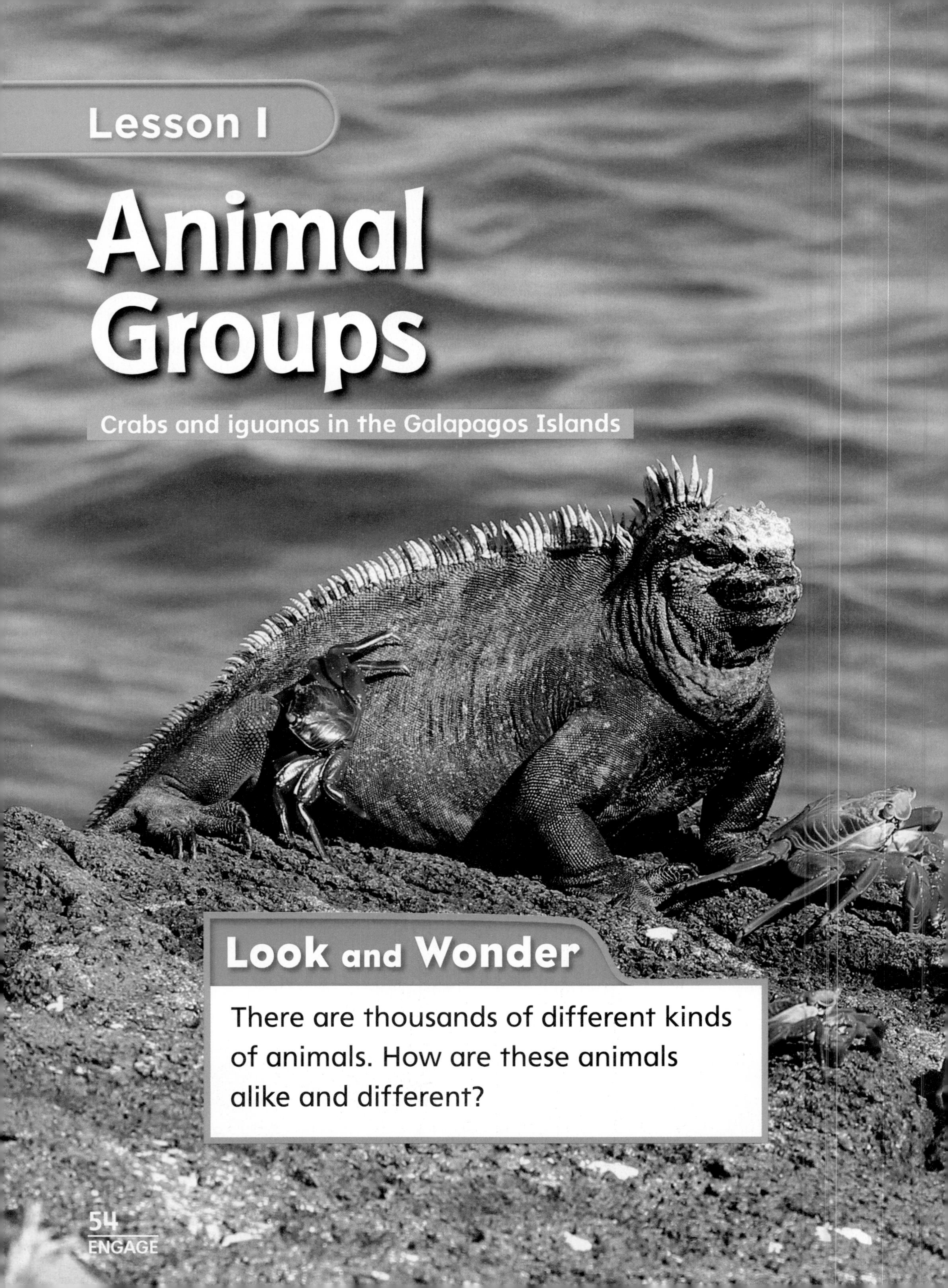

Crabs and iguanas in the Galapagos Islands

Look and Wonder

There are thousands of different kinds of animals. How are these animals alike and different?

How can we put animals into groups?

What to Do

① **Classify.** Look at the pictures of the animals. Put the animals into groups. How did you decide to group the animals?

② Talk about the animal groups with a partner. What groups did your partner use?

③ **Compare.** How are your groups and your partner's groups alike? How are they different?

Explore More

④ **Classify.** Think about animals that live on land. How can you classify them?

How do we group animals?

All animals need food, water, air, shelter, and space to live. They have different parts that help them get what they need to live.

Scientists classify animals into two main groups. One group has backbones. The other group does not have backbones. Here are some animals with backbones.

These lions are mammals. A mammal is an animal that has hair or fur. A female mammal makes milk for her babies. Mammals breathe through their lungs. ▼

This is a bluebird. Birds are the only animals with feathers. All birds have two wings and a beak to help them get food. They lay eggs to hatch their young. ▶

▲ Fish, such as this salmon, live in water. Their gills help them breathe. Their fins help them swim.

▶ This salamander is an amphibian. Most amphibians begin their lives in water. Their moist skin helps them live on land and in water.

This baby alligator is a reptile. It has rough, scaly skin to help protect it. ▼

 Why is a lion a mammal?

FACT ▶ Birds are not the only animals that hatch from eggs. Other animals such as alligators, butterflies, and snakes do, too!

What are some animals without backbones?

There are many kinds of animals that have no backbones. There are more without backbones than with backbones! Some animals without backbones have hard body coverings to help them stay safe.

These jellyfish have no hard body coverings. They sting other animals to stay safe and get food.

Beetle

An insect is an animal with six legs, antennae, and a hard, outer shell.

The antenna helps insects feel, taste, and smell.

The outer shell helps keep insects safe. The legs help insects climb on smooth or rough places.

Read a Diagram

How do the body parts of a beetle help it meet its needs?

Quick Lab

Make a **model** of an animal. Talk with a partner about how the animal meets its needs.

◀ The dragonfly has a hard body covering. It uses its wings to fly away from its enemies.

✓ How do animals without backbones stay safe?

blue crayfish

Think, Talk, and Write

1. **Classify.** How can you classify a lion and a salamander?

2. What do animals need to stay alive?

 3. Choose one animal. Write about a body part from that animal. Describe how it helps the animal meet its needs.

Social Studies L*ink

Make a collage of other animals without backbones. Find out where they live.

LOG ON **e-Review** Summaries and quizzes online at **www.macmillanmh.com**

earthworm

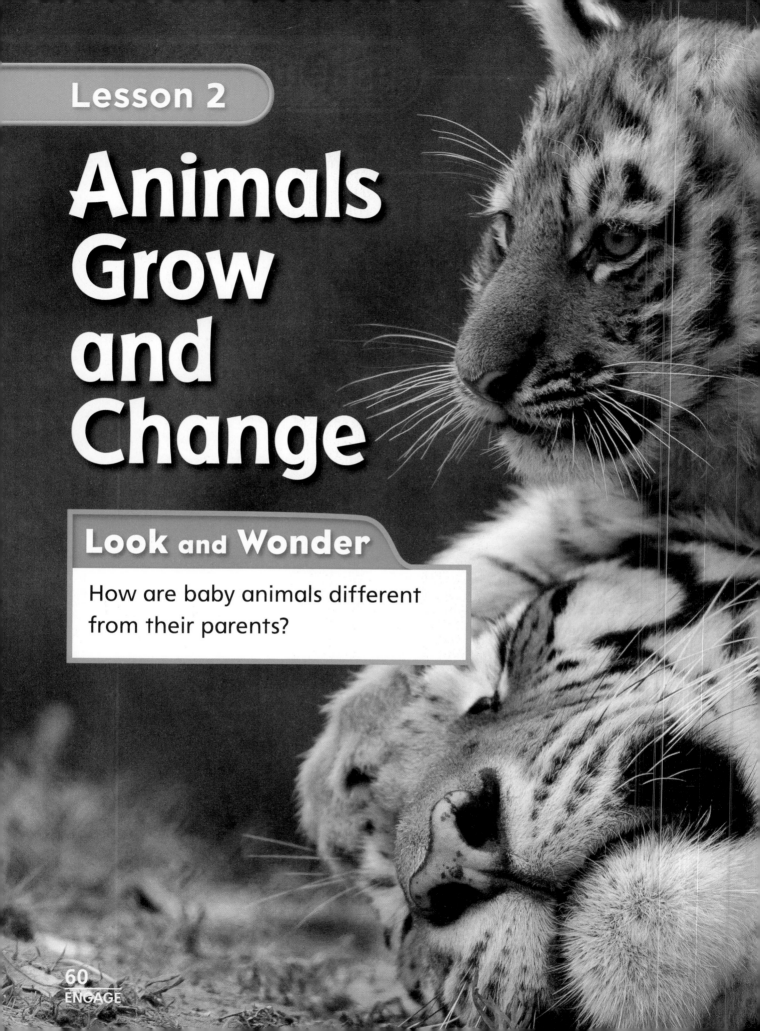

Animals Grow and Change

Look and Wonder

How are baby animals different from their parents?

How are babies and adults alike and different?

What to Do

① What are some things that babies do?

② What are some things adults do?

③ **Compare.** Make a Venn diagram to compare babies to adults.

Explore More

④ How are baby humans and baby tigers alike and different?

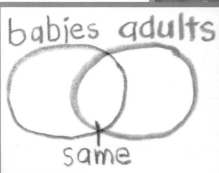

Vocabulary

life cycle

larva

pupa

 SCIENCE QUEST Explore animal life cycles with the Treasure Hunters.

What is a life cycle?

Insects, birds, fish, reptiles, and amphibians lay eggs. Mammals give birth to live babies. Chickens are birds and they lay eggs. All animals have a life cycle. A **life cycle** tells how an animal starts life, grows to be an adult, has young, and dies.

Giant Panda Life Cycle

Baby pandas grow inside their mothers' bodies. They drink milk from their mothers so they can grow.

Chicken Life Cycle

Baby chickens, or chicks, break the shell to get out of an egg. They can see, walk, and feed themselves after they hatch.

Communicate.

Act out a life cycle of an animal.

✓ **What are the stages of a life cycle?**

A baby panda grows up to be an adult. It may find a mate and have a baby of its own.

The chicks grow up to be adult chickens. This is a rooster, or a male chicken.

Read a Diagram

How is the life cycle of a panda different from the life cycle of a chicken?

LOG ON *Science in Motion* Watch animals grow at **www.macmillanmh.com**

FACT It takes 9 months for baby humans to grow before they are born. It takes 4 months for a baby panda to grow.

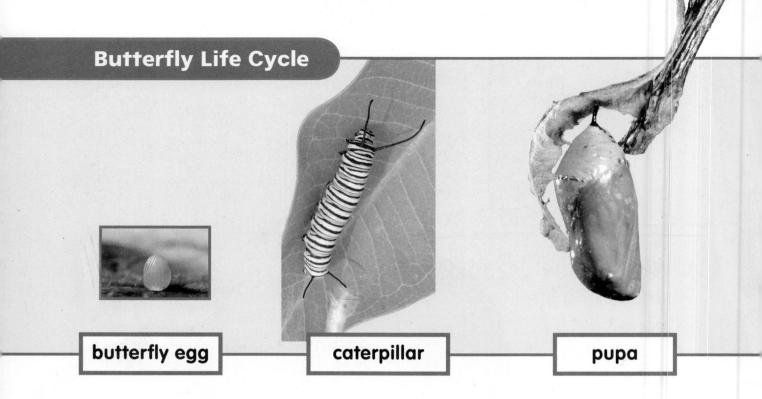

butterfly egg | **caterpillar** | **pupa**

What are some other animal life cycles?

Animals such as butterflies, frogs, and crabs do not start out looking like their parents. They change during their lives.

Butterflies begin as eggs. The next stage after an egg hatches is called the **larva**. A caterpillar is the larva of the butterfly. Caterpillars eat plants to grow.

When a caterpillar is ready to change, it stops moving. Its skin becomes a hard shell. Inside the shell, the caterpillar is slowly changing. This is the **pupa** stage. Soon a butterfly comes out of the shell.

| young butterfly | adult butterfly |

 How does a caterpillar become a butterfly?

Think, Talk, and Write

1. **Predict.** What will the butterfly do when it is an adult?

2. How is the life cycle of a panda the same as the life cycle of a human?

3. Write and draw an example of a life cycle.

Social Studies Link

Research how long five different animals live. Make a chart to put them in order.

LOG ON e-Review Summaries and quizzes online at **www.macmillanmh.com**

Meet Nancy Simmons

Nancy Simmons is a scientist at the American Museum of Natural History. She studies bats all around the world. She has found more than 80 different kinds of bats in one forest. Nancy learns about what bats eat and where they live.

Nancy Simmons is holding a false vampire bat. It is one of the largest bats in the world.

Bats give birth to one baby at a time. The baby is called a pup. The pup is small and pink. It has no hair. To stay safe, the pup hangs on to its mother. The pup gets milk from its mother and grows bigger. After a few months the pup is ready to fly.

Soon the young bat leaves its mother. It can find its own food and start its own family.

Bats hang upside down.

Talk About It

Predict. What will happen to a hairless bat pup as it grows?

AMERICAN MUSEUM OF NATURAL HISTORY

Staying Alive

Look and Wonder

This chameleon searches for food every day. How can it keep from being food for other animals?

How does the color of an animal keep it safe?

What to Do

1 Cut one piece of patterned paper into eight shapes.

2 Put the eight shapes on the other sheet of patterned paper.

3 Time your partner while he or she picks up the shapes.

4 Now put the shapes on plain paper and time your partner again.

5 Which was easier to find? Which was faster? Why?

Explore More

6 **Infer.** How would the activity be different if the shapes were placed on solid colored paper?

You need

scissors

2 pieces of patterned paper

stopwatch

plain paper

Step **1**

Vocabulary

adaptation

camouflage

Why do animals act and look the way they do?

Animals have adaptations to help them stay alive. An **adaptation** is a body part or a way an animal acts that helps it stay alive.

◀ Giraffes have long necks that help them reach leaves in the tops of trees.

A tarsier has big eyes to see at night and long fingers to dig for food. ▶

An anteater can reach insects underground with its long snout. ▶

Camouflage is a way that animals blend into their surroundings. The color or shape of an animal helps it hide. Camouflage keeps animals from being seen by their enemies.

Ptarmigan Feathers

In summer, a ptarmigan has brown feathers.

In fall, the bird's feathers begin to turn white.

In winter, its feathers blend with the snow.

Read a Photo

Why does a ptarmigan turn white in winter?

✓ **What helps animals stay alive?**

The pattern of a snow leopard is hard to see against the rocks.

How do animals stay safe?

There are many different ways that animals act to stay safe. Some animals stay in large groups. Others leave their homes in winter to be in a warm place and to find food.

▲ Sandhill cranes fly south for the winter.

◀ Some animals, like this dormouse, sleep during the cold winter.

Swimming in a large group helps protect these fish from getting eaten by bigger fish.

Animals have body parts to keep them safe. Some animals have shells or smells to protect them from other animals.

 What are some ways animals protect themselves?

Quick Lab

Investigate to find out why eyes are where they are on different animals.

◀ **Turtles stay safe by hiding in their shells.**

Think, Talk, and Write

1. **Cause and Effect.** How does the white fur of a polar bear help it stay alive?

2. Why is it helpful for fish to stay in a group?

3. Write about one adaptation of an animal that keeps it safe.

Health Link

Draw pictures or talk about how you stay safe.

LOG ON **e-Review** Summaries and quizzes online at www.macmillanmh.com

Skunks spray a bad smelling liquid to keep other animals away.

Helpful Traits

Animals have traits that help them live in their environments. Ants have powerful jaws that help them bite and carry food. Frogs have strong legs that help them swim and hop.

angler fish

hummingbird

Write About It

Describe one of the animals above. Where does it live? What do you think it eats? What traits help it live in its environment?

Remember

When you describe, you give details about something.

LOG ON e–**Journal** Write about it online at **www.macmillanmh.com**

Parts of a Group

A dog had 6 puppies. Even though the puppies share many traits, they look different from each other. In this family, 3 of the 6 puppies are brown. You can write this as the fraction $\frac{3}{6}$.

Write Fractions

How many of the 6 puppies are black? Write a fraction to show your answer.

Now draw a group of 3 puppies. Make one third of the group brown.

Remember

You can use a fraction to tell about parts of a group.

So Many Animals!

There are so many kinds of animals! They are alike in many ways. They need food, water, air, and a place to live. Their body parts help them get what they need to stay alive.

Animals are different in many ways, too. Fish have fins to swim and gills to breathe. Birds have feathers to keep warm. Mammals have hair on their bodies and breathe with lungs.

Animals grow in many ways. Some lay eggs. Some give birth to live babies. All of them will grow to look like their parents.

Animals have many ways to keep safe. Some move together in big groups. Others use their colors or shapes to help hide. In the animal world, keeping safe means staying alive!

Vocabulary

Use each word once for items 1–5.

adaptation
amphibian
larva
life cycle
mammal

1. An animal that lives the first part of its life in water and another part on land is an _____.

2. An animal that feeds milk to its young is a _____.

3. How an animal grows and changes is called its _____.

4. This caterpillar is a _____.

5. The blubber of a whale keeps it warm. An _____ like this helps an animal stay alive where it lives.

Answer the questions below.

6. Classify. How would you classify these two animals? List their traits.

7. Predict. What will happen when a chick hatches from an egg?

8. Put these pictures of a frog life cycle in order.

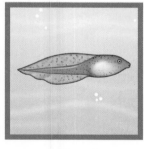

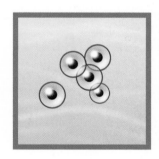

| tadpole | adult frog | eggs | tadpole with legs |

9. What are some ways animals can keep safe?

The Big Idea

10. How do animals grow and change?

Careers in Science

Bird Bander

Do you love to learn about birds? You could become a bird bander. A bird bander helps scientists keep track of birds.

The bander catches a bird and puts a tiny band around its ankle. This band has a number on it, and the bander writes it down. The bander also writes the bird's age and size.

Then the bander returns the bird to the wild. Later, other banders and scientists might trap the same bird. They can look up the bird's number and see how it grew and changed.

bird bander

More Careers to Think About

wildlife guide

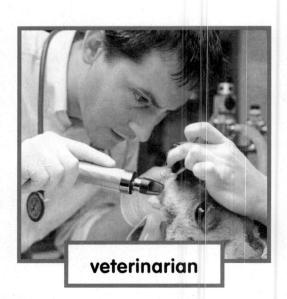

veterinarian

Habitats

Monitor lizards can stand up on two legs and hiss.

Little Sandy Desert, Australia

In Payment

Caterpillars like to crunch
On stems and roots and leaves
Until the plant has many holes
That make it like swiss cheese

Then the caterpillar spins its silk
To make its own cocoon
Turning into a butterfly
Is what it will do soon!

The butterfly will then take pollen
To flowers of red and yellow,
Which makes up for the nibbling
Now that he is a helpful fellow!

Talk About It

Where have you
seen a caterpillar?

CHAPTER 3

Looking at Habitats

The Big Idea

What are habitats?

Beavers making a dam in Wyoming

Key Vocabulary

habitat a place where plants and animals live

(page 90)

predator an animal that hunts other animals for food

(page 97)

drought a long period of time with little or no rain

(page 104)

fossil what is left of a living thing from the past

(page 108)

Places to Live

Reef fish in the Bahamas

Look and Wonder

What can you tell about the place these plants and animals live?

Where do animals live?

You need

paper

crayons

What to Do

1. **Observe.** Look at the footprints below. What animal do you think made them?

2. **Infer.** How does the shape of its feet help this animal? Share your idea with a partner.

3. Draw a picture of the animal and the place where it lives.

Explore More

4. **Communicate.** What other animals could live near this animal? What do they need to live? How do they get food and water? Make a chart.

Vocabulary

habitat

What is a habitat?

A **habitat** is a place where plants and animals live. In a habitat, animals can find the food, homes, and water they need to live. Plants need soil, rain, sunlight, and animals in their habitats to live.

grassy and warm

cold and snowy

wet and grassy

There are many kinds of habitats. Some have lots of rain. Some are dry. Some places are windy and others are cold.

Different plants and animals need different habitats to live. These pictures show some kinds of habitats.

 What are some kinds of habitats?

hot and dry

How do living things use their habitats?

Animals use the plants living in their habitat for food. Some animals eat other animals that live in the same habitat. Animals also use their habitat to hide and sleep.

Animals such as moles dig tunnels in the soil to find food and shelter. Some insects make their homes under rocks.

Forest Habitat

Read a Diagram

How do the squirrel and the snake use their habitat?

Different plants need different kinds of soil to live. Some plants grow in sandy soil and some plants grow in rocky soil.

Plants that live in dry places can hold water. Plants that live in very wet places can get rid of extra water. They have leaves that point down so water can roll off them.

 How do animals and plants use their habitats?

Quick Lab

Find a picture of a habitat. Draw and write to **communicate** what could live there.

This plant lives in a dry place. Its leaves store water. ▶

Think, Talk, and Write

1. **Summarize.** How are habitats different?

2. How do plants survive in their habitat?

3. Write about a hot, dry habitat. Describe what you would find there.

Art Link

Draw a picture of a habitat you want to visit. How would you get what you need there?

Food Chains and Food Webs

Look and Wonder

Animals need food to live.
What do different animals eat?

Animals can eat plants or other animals. An animal that hunts other animals for food is a **predator**. Animals that are hunted by predators are called **prey**.

Some animals eat plants and animals that are dead. Animals such as worms break the dead things up into very small pieces.

Quick Lab

Communicate. Act out a food chain with puppets.

✔ Where can food chains be found?

The snake eats the lizard.

The hawk eats the snake.

Large fish, such as tuna, eat sea horses.

Sharks eat large fish.

A Desert Food Web

Arrows in the food web go from food to eater.

Read a Diagram

What are the different food chains in this food web?

LOG ON *Science in Motion* See the parts of a food web at **www.macmillanmh.com**

What is a food web?

A **food web** is two or more food chains that are connected. Sometimes one kind of animal is food for many animals. Mice are eaten by hawks, owls, and snakes.

Animals also eat more than one kind of animal. Hawks eat mice, rabbits, frogs, and snakes. If you put those food chains together, you have a food web.

The insect is prey for the bird.

✓ What are some other predators and their prey?

Think, Talk, and Write

1. **Main Idea and Details.** Describe an example of a food chain.

2. What is a food web?

3. Write about how you are part of a food chain.

Health Link

Think of a healthy lunch. Show how your meal is part of a food web. Draw the web.

LOG ON e-Review Summaries and quizzes online at www.macmillanmh.com

A Food Web for Lunch

Emma is having a chicken sandwich for lunch. She drew a food web to show how each food is related.

Emma's Food Web

me

chicken

beetle

wheat

lettuce

✏ Write About It

Explain how Emma, the chicken, lettuce, and wheat form a food web. Think about the food chains in Emma's lunch to help you form a food web of your own lunch.

Remember
When you are writing to explain, you tell the steps in order.

LOG ON ⊝ –Journal Write about it online at www.macmillanmh.com

Food for a Toad

Most animals eat different foods to stay alive.

Problem Solving

A toad ate 3 grasshoppers on Monday. It ate 5 ants on Tuesday. It ate 4 crickets on Wednesday. How many animals did the toad eat in all?

Remember

Making a sketch can help solve problems. Think about whether you need to add or subtract.

Habitats Change

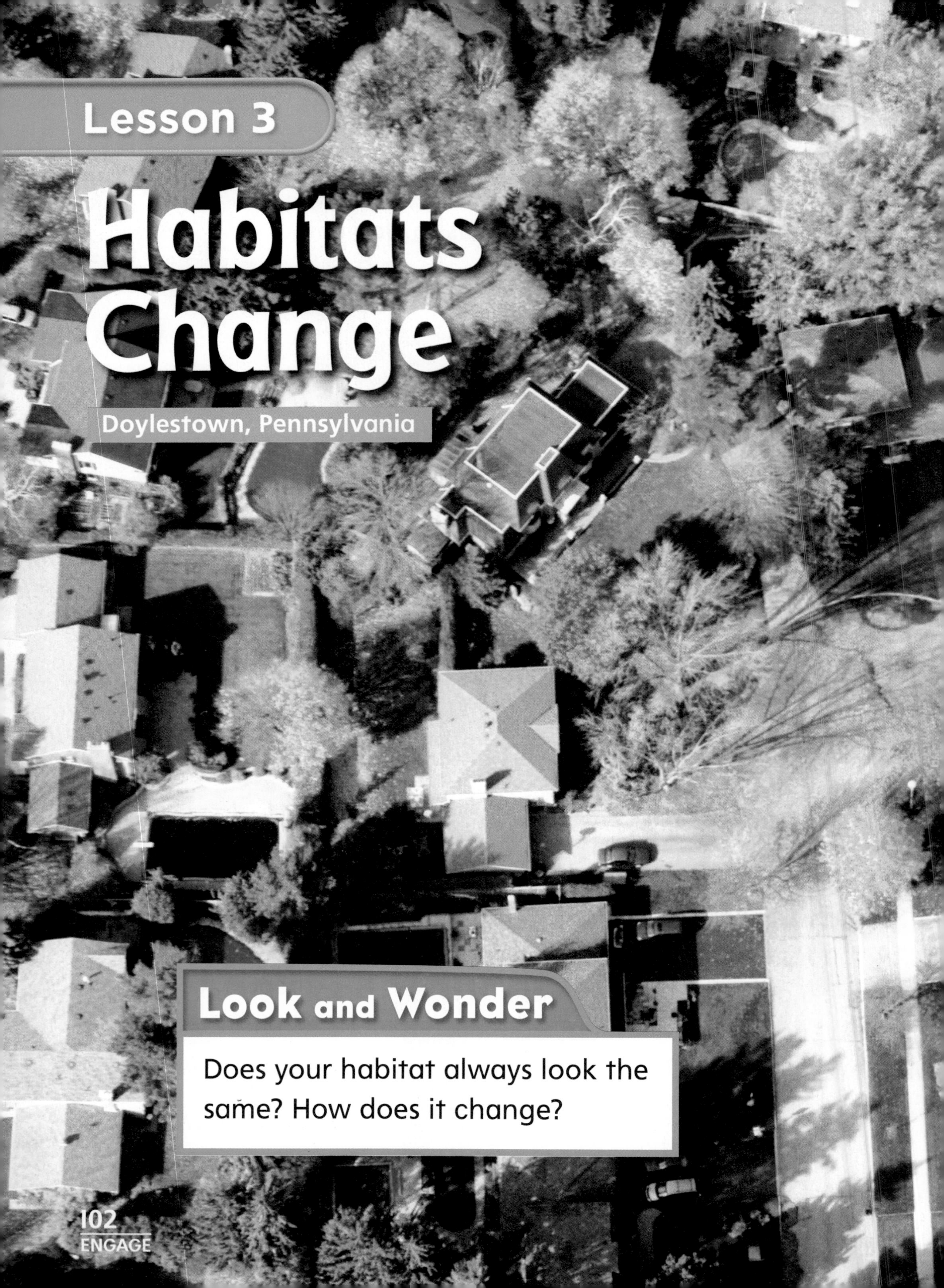

Doylestown, Pennsylvania

Look and Wonder

Does your habitat always look the same? How does it change?

What happens when habitats change?

What to Do

You need

large pieces of paper

crayons

small toys and blocks

1. On a large sheet of paper, draw a large meadow, woods, and river.

2. Place the animals where they would live.

3. Use blocks as houses and buildings. Build a town with houses and stores.

4. **Observe.** What happens to the meadow, woods, and animals that live there?

5. **Infer.** How does building a town affect the animals, meadow, woods, farms, river, and people?

Explore More

6. **Predict.** What will happen if a highway is built?

Step **2**

Vocabulary

drought

endangered

fossil

extinct

SCIENCE QUEST Explore fossils with the Treasure Hunters.

How do habitats change?

Nature can change habitats in many ways. A **drought** is a long period of time when there is little or no rain. Plants and animals can not live without water. Floods or fires can also change habitats.

Drought

Read a Photo

How does a drought change a habitat?

Animals can change habitats. Beavers make dams. The dam can make a pond.

People can change habitats, too. People build houses and other buildings where grass, plants, and trees were growing.

▲ **Bulldozers help people clear land to build on. Clearing the land changes the habitat.**

 How can a habitat change?

The beaver dam blocks the water from the stream and forms a pond.

What happens when habitats change?

When a habitat is changed, animals may not be able to find the things they need. Some animals may die. When many of one kind of animal die and only a few are left, that animal is **endangered**. All these animals are endangered.

Quick Lab

Draw a comic strip about a habitat. **Communicate** how habitats can change.

whooping crane

tigers

People hunt tigers for their fur and cut down their forest homes.

People built over the marshes where cranes live.

Animals can become endangered when people hunt them or build on their habitats.

When habitats change, some animals have adaptations that help them live in their new habitat. Animals may find new places to get food and live.

 Why do animals become endangered?

manatee

People have taken over many rivers where manatees live. Fishing nets and powerboats also hurt manatees.

FACT People helped American alligators survive so they are not endangered anymore.

How can we tell what a habitat used to be like?

Scientists study fossils to learn about Earth's past. A **fossil** is what is left of a living thing from the past. Scientists get clues about habitats of the past from the plant and animal fossils they find.

Some fossils do not match the habitat they were found in. Then scientists can tell that the habitat has changed.

Look at the fossil found here. What do you think the habitat used to be?

How do you think this animal moved? Why?

Some plants and animals that lived long ago still live today. Some have died out, or become **extinct**. Now we only have their fossils. Fossils can help tell how animals may have looked or moved.

 What can fossils tell us about habitats long ago?

Think, Talk, and Write

1. **Cause and Effect.** What happens to plants and animals when their habitat changes?

2. What are some ways animals can stay alive when their habitat changes?

3. Write about how people can change habitats and what might happen to animals and plants.

Art Link

Make a poster about endangered animals.

LOG ON e-Review Summaries and quizzes online at www.macmillanmh.com

Meet Mike Novacek

Mike Novacek grew up in Southern California. When he was young he visited the La Brea Tar Pits in Los Angeles. He loved to learn about the fossils he saw there. He learned about animals that lived long ago.

Mike Novacek

▼ **Mike travels to the Gobi Desert in China to look for new fossils.**

Today Mike is a scientist at the American Museum of Natural History in New York. He travels all around the world to collect fossils. He looks for fossils of reptiles, mammals, and dinosaurs. Many of these animals lived 80 million years ago!

Mike and his team went to the Gobi Desert to look for fossils. They found fossils of the Kryptobaatar. These were tiny mammals the size of a mouse. These mammals lived at the same time and place as dinosaurs!

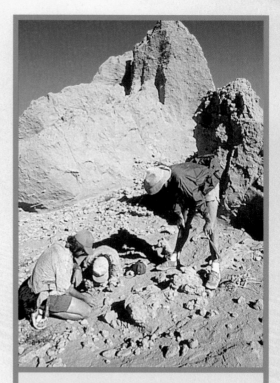

Mike studies fossils to find out about the past.

Talk About It

Cause and Effect. What made Mike want to study fossils?

▲ **Kryptobaatar skull**

AMERICAN
MUSEUM ᴼꜰ
NATURAL
HISTORY

Changing Habitats

This is a woodland forest habitat. Deer, birds, fish, and snakes live here. There are many kinds of plants and trees here, too. What would happen if this habitat changed?

Oh no! A fire is coming! Deer run away. Snakes slither out from under rocks. Birds fly far from the smoke. Most of the animals can stay safe away from the fire.

113

The fire is over. Some plants burned. Some plants survived. New plants like fireweed and grasses grow quickly. These plants have more room to grow after the fire. They can also get more sunlight.

The animals come back and eat the new plants. Deer love the soft new leaves. A snake slithers back under a rock. The birds fly back to the trees that survived. Foxes make a new den.

Vocabulary

Use each word once for items 1–5.

| drought |
| endangered |
| food chain |
| habitat |
| prey |

1. When there are not many of one kind of animal left, the animal is called _____.

2. When it does not rain for a long time there is a _____.

3. A place where animals and plants live together is called a _____.

4. The grasshopper in this picture is the _____.

5. The picture below shows part of a _____.

Answer the questions below.

6. Summarize. How do plants and animals use each other?

7. Compare the pictures below. How are they different? What do you think happened?

8. Put Things in Order. Put this food chain in order.

| rabbit | Sun | fox | grass |

9. What happens to animals and plants when habitats change?

10. What are habitats?

CHAPTER 4

Kinds of Habitats

The Big Idea

What are different kinds of habitats?

Polar bears in Canada

Key Vocabulary

rain forest a habitat where it rains almost every day (page 124)

woodland forest, page 122

desert, page 130

Arctic a very cold place near the North Pole (page 132)

ocean a large body of salty water

(page 136)

pond a small body of fresh water

(page 138)

Forests

Banff National Park, Canada

Look and Wonder

How is this forest like one you know? How is it different?

What is a forest like?

What to Do

① **Make a model** of a forest. Place soil, a plant, and rocks in a bottle.

Step ①

② Water the soil. Add a pill bug. Cover the bottle with plastic wrap. Poke holes in it. Place near a window.

③ **Observe** your model. Record on a chart how it changes.

Explore More

④ **Make a model** of the forest in winter. Draw a picture to show how it would change.

You need

soil

plant

plastic wrap

rocks

plastic spoon

pill bug

Vocabulary

woodland forest

rain forest

What is a woodland forest like?

A **woodland forest** is a habitat that gets enough rain and sunlight for trees to grow well. A habitat is a place where plants and animals live. Deer, black bears, and foxes live in woodland forests.

Birds, insects, and worms live there, too. Most trees have leaves that turn color and drop to the ground in the fall. There are some trees that stay green all year.

Life on a Forest Floor

woodchuck

white-tailed deer

opossum

Forest animals survive in different ways. Bears eat berries, nuts, and fish in the spring, summer, and fall. They sleep in caves or hollow logs in the winter when there is not much food around.

 chickadees

Animal Traits

 Owls have large eyes and hunt at night using sight and sound.

 Woodpeckers have sharp beaks to tap trees and find insects to eat.

 Deer have colors and spots that help them hide in the forest.

 Chipmunks have big cheeks that they use to carry many nuts.

Read a Chart

How does the woodpecker get food?

raccoon

✓ How do some animals survive in the woodland forest?

What is a tropical rain forest?

A tropical **rain forest** is a habitat where it rains almost every day.

Many kinds of plants and animals live in different parts of the rain forest. Bats, insects, colorful birds, and other animals live in the treetops. Jaguars, tapirs, and wild boars are some of the animals that live on the ground.

Quick Lab

Observe worms and record what you see.

▶ **This parrot eats fruit and seeds from high trees.**

▼ **This tapir hunts for its food on the ground.**

Trees grow tall and have huge leaves. The treetops block much of the sunlight. Orchids, ferns, and mosses live on trees to get enough sunlight.

These plants never touch the ground. Animals and plants try to blend in with the trees to stay safe.

 How do plants survive in the rain forest?

The roots on this orchid collect water from the air and rain.

Think, Talk, and Write

1. **Compare and Contrast.** How are a rain forest and a woodland forest alike? How are they different?

2. Name animals that live in the rain forest.

3. Draw and write about how animals stay safe in a woodland forest.

Art Link

Draw a picture of an animal in the rain forest. Show how it is protected by its color or shape.

Some animals have colors and patterns that scare away other animals.

LOG ON e-Review Summaries and quizzes online at www.macmillanmh.com

Meet Liliana Dávalos

Liliana Dávalos is a biologist. That is a scientist who studies living things.

Many kinds of animals live in the Amazon rain forest of South America. Some of these animals are losing their homes because trees are being cut down.

Liliana Dávalos is a scientist at the American Museum of Natural History. She is worried about these animals. She is concerned about tiny singing birds called manakins. These birds live in Colombia. They are endangered.

This is the wire-tailed manakin. It is one of the most colorful manakins of the Amazon rain forest.

Liliana works to help protect the manakins and their homes. She counts the manakins and other birds. She also takes pictures of them. Then she records this information on a map. The map can help people decide which forests and animal homes to protect.

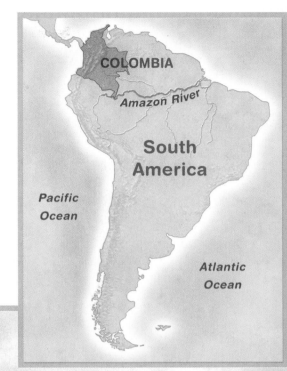

People cut down this forest to sell the trees for their wood.

Talk About It

Compare and Contrast. What other habitats have you learned about? How are they like or different from the rain forest?

AMERICAN
MUSEUM OF
NATURAL
HISTORY

127
EXTEND

Hot and Cold Deserts

Desert wildflowers in California

Look and Wonder

How do you think these plants stay alive?

How does the shape of a leaf help a plant?

What to Do

You need

paper
towels

1. Cut two leaf shapes from paper towels.

2. Roll up one leaf shape and tape it closed.

Step 2

scissors

tape

3. Place both leaf shapes on plastic wrap and wet them.

4. **Observe.** Check both leaf shapes every 15 minutes. Which leaf shape stayed wet longer?

water

Explore More

5. **Draw Conclusions.** Which kind of leaf shape might you find in a dry place?

plastic wrap

What is a hot desert like?

A **desert** is a dry habitat that gets very little rain. Hot deserts are hot during the day and cool at night. The soil there is sandy and rocky. Cactuses and grasses grow in hot deserts.

Desert plants can store water in their stems and leaves. Some have roots that spread out far or deep to find water when it rains. Some leaves curl up during the day to hide from the sunlight.

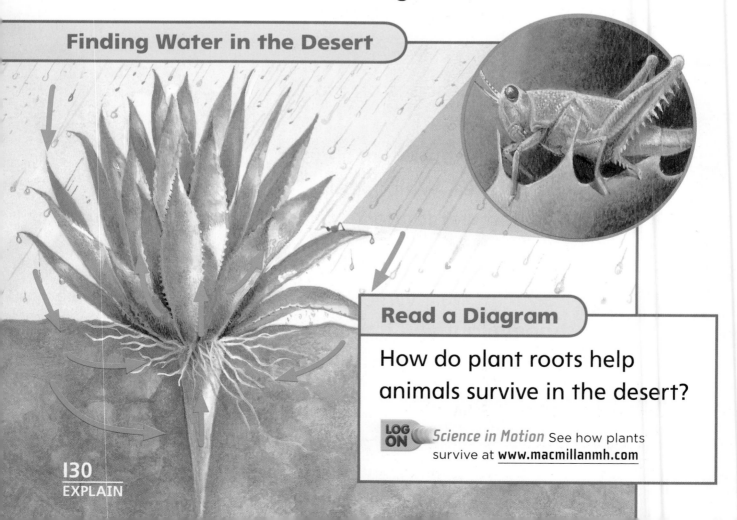

Finding Water in the Desert

Read a Diagram

How do plant roots help animals survive in the desert?

LOG ON *Science in Motion* See how plants survive at **www.macmillanmh.com**

snake

Gila monster

Tortoises, snakes, and lizards can live in the desert. Things that live in the desert need to survive without much water.

Desert animals get most of their water from eating plants or other animals. Many animals have pale colors that help them stay cool and hide from other animals. Most animals sleep during the day and come out at night when it is cooler.

horned lizard

coyote

How do animals survive in a desert habitat?

What is the Arctic like?

There are cold deserts, too. The **Arctic** is a desert near the North Pole. It is cold and dry. Arctic foxes, reindeer, polar bears, and walruses live there. Polar bears and foxes grow thick white fur to keep them warm and help them hide. Thick blubber helps walruses and seals survive.

Quick Lab

Investigate to find out how animals survive in the Arctic.

Walruses rest on an ice floe.

▼ ermine

reindeer ▶

In the Arctic there are many small, low plants but no tall trees. Arctic plants grow close to the ground, sheltered from cold winds.

These plants have tiny leaves. The roots are shallow because the soil under the surface stays frozen all year.

 How do plants and animals survive in the Arctic?

Think, Talk, and Write

1. **Classify.** Make a list of hot desert animals and cold desert animals.

2. What are some plants that live in hot deserts?

3. Draw and write about how the hot desert and the Arctic are the same and different.

Art Link

Draw a picture of animals in the desert and the Arctic. Show how they survive in these harsh climates.

FACT The growing season in the Arctic is so short that most plants bloom at the same time.

Oceans and Ponds

Look and Wonder

How do animals and plants survive in saltwater habitats?

What lives in a saltwater habitat?

What to Do

1 Fill each container with two cups of water. Add two teaspoons of salt to one container and stir.

2 Add $\frac{1}{4}$ teaspoon of brine shrimp eggs to each container.

Step **2**

3 **Observe** with a hand lens what happens every day for a week. Record your results in a table.

Explore More

4 **Record Data.** What happens if you place the shrimp in a dark place for a week?

You need

2 containers

salt

measuring spoons

measuring cup

brine shrimp eggs

hand lens

What is the ocean like?

The **ocean** is a large body of salt water. The water is always flowing. The ocean covers most of Earth.

Kelp is a kind of ocean plant, or seaweed. Beds of kelp are like forests in the ocean. They provide food and shelter for many kinds of ocean animals.

▼ **Different animals live in different parts of the ocean.**

Life in the Ocean

Whales, dolphins, and sharks swim in the deep water, away from shore.

Crabs and sea stars live closer to the ocean's shore.

Read a Diagram

What animals live in the deep ocean?

Most animals that live in the sea must move through the water. Their body shapes and fins or flippers help them swim. Jellyfish and squid move by sucking in water and forcing it out.

Ocean animals also must stay safe. Sea turtles and clams are protected by shells. Other animals have spines.

✓ How do living things survive in the ocean?

Quick Lab

Investigate. Use a bottle of water in a tub of water to see how fish move through water.

Sea stars and blowfish have spines to protect them.

This shark is shaped like a rocket to move fast.

Jellyfish sting.

Angler fish and squid live in the deepest part of the ocean.

Sea turtles use their flippers to swim.

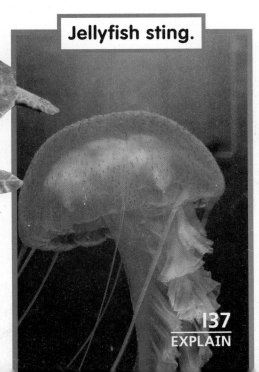

What is a pond like?

A **pond** is a small body of fresh water that does not flow. Fresh water has little or no salt in it. Most animals that live in fresh water can not live in salt water.

Frogs, fish, and turtles eat insects that live in ponds. Snakes live in the grass near ponds and eat fish and frogs.

little blue heron

water snake

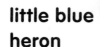
fish

muskrat

Pond plants with roots grow near the shore. The stems, leaves, and flowers grow out of the water to get sunlight.

Pond animals breathe in different ways. Some insects, such as mosquito larvae, send tubes to the surface. These insects breathe through the tubes. Salamanders breathe through their skin. Fish breathe through gills.

The whirligig beetle carries a bubble of air under water to breathe.

✓ How do living things breathe in a pond habitat?

Think, Talk, and Write

1. **Main Idea and Details.** What plants and animals live in the ocean?

2. How do pond plants grow?

 3. Draw and write about how a pond and an ocean are the same and different.

Social Studies Link*

Read about an ocean far from where you live to see what lives in that habitat.

A Visit to the Ocean

What would it be like to visit an ocean?

✎ Write About It

Write a story about a trip you might take to the ocean. How would you get there? Who would you go with?

Describe in your story what you would see, hear, and do. Write about how it might feel to be there.

LOG ON ⒠ –Journal Write about it online at **www.macmillanmh.com**

Remember

A story has a beginning, middle, and end. A story uses describing words.

A Pond Bar Graph

Jim visited a pond with his family.
Jim saw many different animals.

salamander

fish

frog

turtle

Make a Bar Graph

Jim saw 8 salamanders, 6 frogs, 10 fish, and 3 turtles. Make a bar graph like the one shown. Record how many of each animal Jim saw at the pond.

Remember
Use a ruler to make straight lines. This will make your graph easier to read.

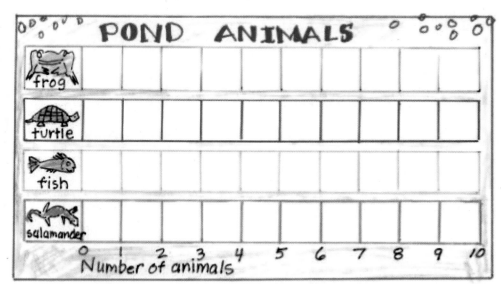

POND ANIMALS

frog

turtle

fish

salamander

0 1 2 3 4 5 6 7 8 9 10
Number of animals

Different Habitats

A woodland forest is one kind of habitat. Bears and mice live in woodland forests. Most trees have leaves that change colors when seasons change. Some other plants stay green all year.

Some animals live in cold places such as the Arctic. The Arctic is a kind of desert. Musk oxen, polar bears, and reindeer live there. They all have different ways to survive the cold winters.

A coral reef is a kind of living ocean habitat. Coral is made up of tiny animals. Fish and other animals swim in the coral reef. They can hide and find food there.

A pond is another kind of water habitat. Alligators live in some ponds. The alligator can hide under the water to catch food. Fish and other animals can hide among the roots of the trees.

Vocabulary

Use each term once for items 1–5.

Arctic

desert

ocean

rain forest

woodland forest

1. This hot, dry habitat is called a _____.

2. This cold habitat is the _____.

3. A habitat like this is called a _____.

4. This giant body of salt water is an _____.

5. This tropical habitat is a _____.

Answer the questions below.

6. Infer. What kind of habitat is this?
What animals and plants live here?

7. Compare and Contrast. How does a hot
desert compare to an Arctic habitat?

8. How do animals survive in very different
kinds of habitats?

9. What are different kinds of habitats?

Marine Mammal Scientist

Do you want to help protect dolphins and whales? You could become a marine mammal scientist.

Marine mammal scientists protect habitats in many ways. Some study the oceans and the animals that live there. They learn more about what animals need. Some teach people why it is important to take care of the oceans and the animals.

Some other jobs for people who want to protect marine animals are veterinarian and underwater photographer.

marine mammal scientist

More Careers to Think About

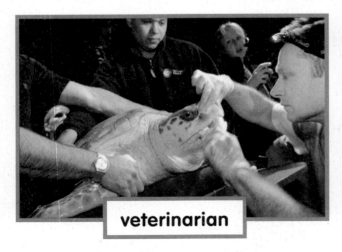

veterinarian

underwater photographer

Our Earth

The rocks of Bryce Canyon are called hoodoos.

Earthworms
Soil Helpers

You live your life on top of Earth. At the same time, many things are living under your feet. Many animals are busy in the ground.

An earthworm is one animal that lives underground. They live in soil.

They don't have hands or feet. They don't have arms or legs. Yet they are able to move underground. They move by eating almost everything in their path. But they do that without any teeth!

Earthworms have an important job. They eat the bits of dirt and dead plants or animals that they find. Their bodies make these bits into material that helps the soil. They make the soil better.

As they move, the help mix the soil. This helps the plants that grow in soil. The plants roots are able to grow and spread. Earthworms are important soil helpers.

Talk About It

How do earthworms help the soil?

CHAPTER 5

Land and Water

The Big Idea How can we describe Earth's land and water?

Point Lobos State Park, in California

Key Vocabulary

landform one of the different shapes of Earth's land (page 156)

earthquake a shake in Earth's crust (page 174)

volcano an opening in Earth's mantle and crust (page 174)

flood water that flows over land and can not easily soak into the ground (page 175)

Earth's Land

Pyrenees Mountains, Spain

Look and Wonder

How is this place like where you live? How is it different?

How are parts of Earth's surface alike and different?

What to Do

1. **Observe.** How are the pictures alike? How are they different? Talk about the pictures with your partner.

2. **Classify.** Sort the pictures into two groups. Describe how you sorted them.

3. **Classify.** This time sort the pictures into three new groups.

Explore More

4. **Predict.** How might the land change during a year?

What is land like on Earth?

What is the land like where you are? Land can be smooth, rocky, or flat. Land can be high or low. Earth's land has different shapes. Each of these shapes is called a **landform**.

A mountain is a high area of land. Mountains may have pointed tops and steep sides.

A valley is low land between mountains or hills.

The land is different all over Earth. Some areas have a lot of mountains and valleys. Other areas have flat plains.

A plain is an area that is flat and wide. Plains cover many miles.

Hills are small mounds of Earth. They are not as high as mountains.

 How is a mountain different from a valley?

What can maps tell us about Earth?

Have you ever looked at a map of Earth? A globe is a model of Earth. It is a map in the shape of a sphere, or ball. It shows where land and water are found.

Each big piece of land is a continent. Continents are surrounded by oceans. Small pieces of land with water on all sides are islands.

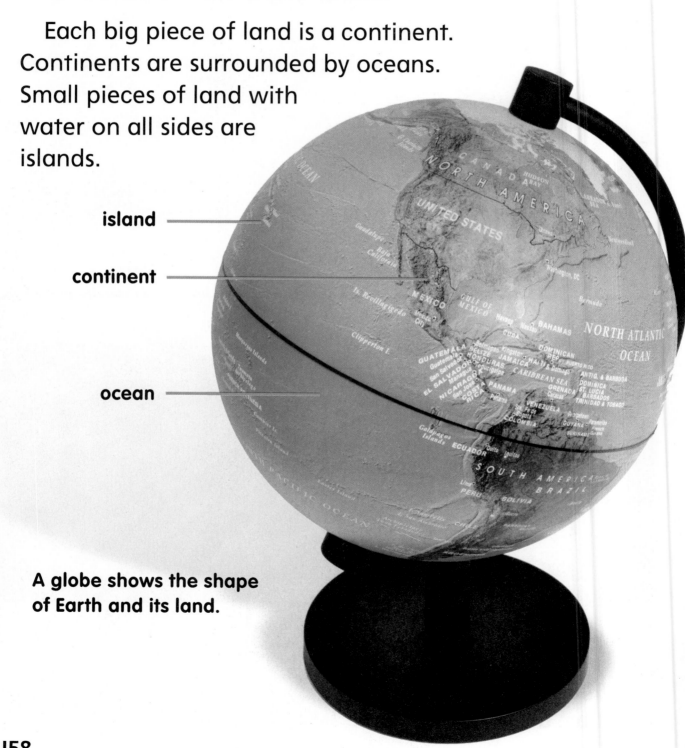

island

continent

ocean

A globe shows the shape of Earth and its land.

Maps can also be flat. The map on this page shows the continent we live on.

Some maps show where the land is high or low. Look at the colors on the map. Brown shows mountains. Green shows flatter land.

 What can we learn about Earth's land by looking at maps?

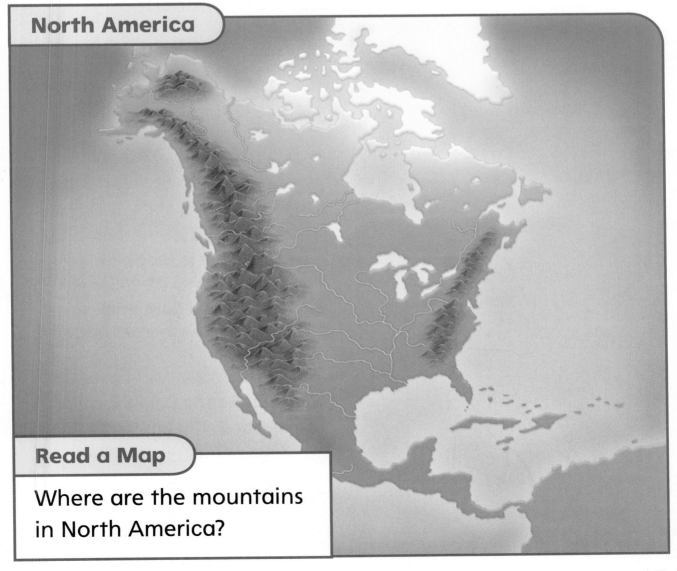

North America

Read a Map

Where are the mountains in North America?

What is inside Earth?

What is Earth like far below the ground? Earth is made of three main layers. The part of Earth that we live on is called the **crust**. It is the outer layer of Earth. Earth's crust is very hard. It is the thinnest layer.

Below the crust is a very hot layer called the **mantle**. The center of Earth is the **core**. Part of the core is liquid. Part of the core is solid. The core is the hottest part of Earth.

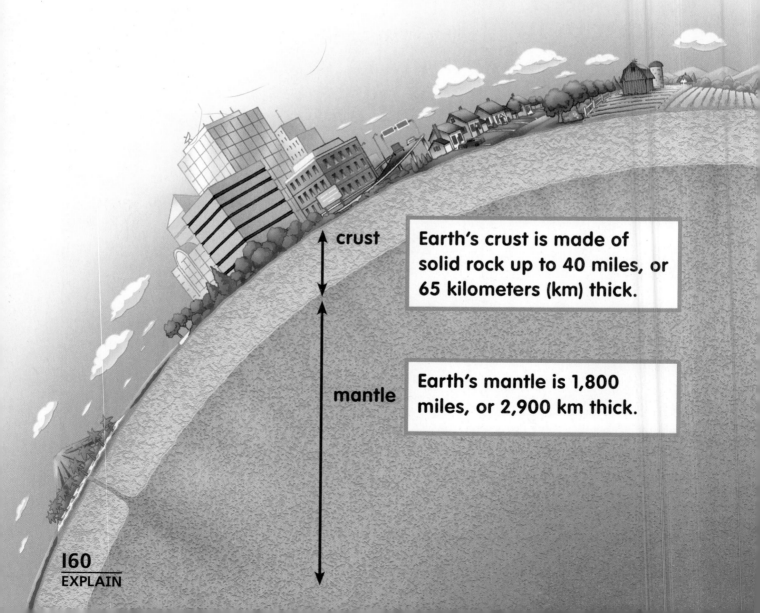

crust

Earth's crust is made of solid rock up to 40 miles, or 65 kilometers (km) thick.

mantle

Earth's mantle is 1,800 miles, or 2,900 km thick.

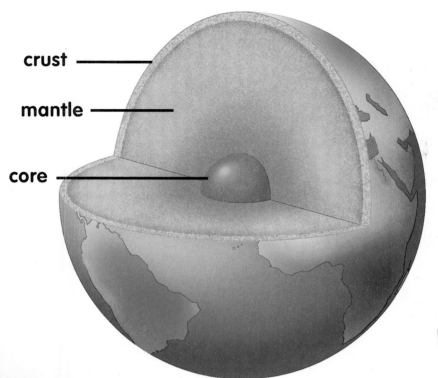

crust

mantle

core

Quick Lab

Make a model of Earth's layers. Use colors of clay to show the core, mantle, and crust.

 How is Earth's crust different from the core?

Think, Talk, and Write

1. **Compare and Contrast.** How are continents different from islands?

2. How are hills different from mountains?

3. Write about the layers of Earth.

Social Studies Link*

Make a map of your classroom. Use symbols to show different objects around the room.

LOG ON e-Review Summaries and quizzes online at **www.macmillanmh.com**

Earth's Water

Honaunau, Hawaii

Look and Wonder

How are these children using water?

How do people use water?

What to Do

1. **Observe.** How many ways do you use water during the day? Make a list.

2. **Communicate.** Discuss your list with classmates and add other ways that people use water.

3. **Classify.** Sort the different ways people use water. How can you classify the different ways?

Explore More

4. **Investigate.** How many times do you use water in one day? Make a tally chart. Use your tally to make a bar graph.

How People Use Water	
drinking water	III
washing dishes	
swimming	

Why is Earth's water important?

Think about what your day would be like without water. You could not drink or wash your hands. The dishes you use to eat would stay dirty. We need water every day.

Living things on Earth can not survive without water. Plants use water to stay healthy and make their own food. Humans and many other animals need fresh water to drink.

▶ **Flowers depend on rain for water.**

▼ **These lions are drinking from a freshwater stream.**

Fresh water is water that is not salty. Where is fresh water found on Earth? Water is found in lakes, ponds, rivers, and streams.

Ice and snow melt and flow down hills or mountains. Then this water enters streams and rivers. Some of the water is then cleaned for people to use.

This dam helps supply fresh water to a city or town. The water is cleaned. Pipes then bring the water into our homes. ▶

What are some places fresh water is found on Earth?

Where is most of Earth's water found?

Have you ever been to the ocean? An **ocean** is a large, deep body of salty water. The ocean looks as if it goes on forever.

Oceans cover about three-fourths of Earth. Think of Earth as four equal parts. Oceans cover three of those four parts.

Earth from Space

Read a Photo

What do the blue parts on Earth show?

FACT You should not drink salt water.

People have used oceans for hundreds of years. Ships carry people and goods around the world.

Many living things live in the ocean. These living things need salt water to stay alive.

▲ These ships travel on the ocean.

✔ Why do you think people do not drink ocean water?

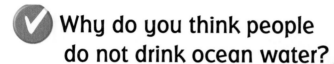

◀ Most dolphins live in salt water.

Think, Talk, and Write

1. **Summarize.** How much of Earth is covered by ocean?

2. How do plants and animals use water?

3. Write about how living things use the ocean.

Art Link

Make a collage of why water is important.

LOG ON ℮-Review Summaries and quizzes online at **www.macmillanmh.com**

My Water

Where can you find bodies of water where you live? Learn more about the water in your town or state from a book or from the Internet.

Write About It

Write a report about lakes, streams, or ponds where you live. Tell what animals live there. Tell how you can help protect them and the water. Draw a picture of the water. Share your report with the class.

Remember
A report gives many details about a topic.

LOG ON e–**Journal** Write about it online at **www.macmillanmh.com**

How Much Water?

After it rains, ponds and lakes can be deeper than before.

Choose the Operation

Yesterday this pond was 18 inches deep. It rained last night. Now the pond is 6 inches deeper. How deep is the pond after it rained?

Remember

Think about what you want to know. Do you want to add or subtract?

Changes on Earth

Nova Scotia, Canada

Look and Wonder

Crash! Waves hit these rocks hard every day. How do you think water can change rocks?

How can you change rocks?

What to Do

① **Observe.** Use a hand lens to look at three or four rocks. Draw what you see.

② **Predict.** What will happen if you shake the rocks in a jar with water?

③ Put the rocks in a jar filled half way with water. Shake the jar for two minutes. Remove the rocks. Observe them again.

④ **Draw conclusions.** Why do you think the rocks changed or did not change?

Explore More

⑤ **Experiment.** How else can you change the rocks?

You need

rocks

jar of water

hand lens

Step ③

Vocabulary

earthquake

volcano

flood

landslide

◀ **The river cut through this rock.**

How does Earth change slowly?

Earth changes every day. You can not see some of these changes. Many changes take thousands of years.

Water changes the land over time. Fast moving water can wear away rock.

Strong winds blow sand and carry away soil. The surface of the land or rock changes shape.

Wind shaped these sand dunes over time.

FACT Not all sand is found at the beach.

Ice can shape the land, too. Over many years ice can break large rocks into little pieces.

 How can wind and water change rocks?

≡Quick Lab

Fill a small plastic bottle with water and close tightly. **Predict** how the bottle will change in the freezer overnight.

Ice Changes Land

1

In cold weather, the water turns to ice. Ice takes up more space than water. Ice pushes the cracks apart.

2

In warmer weather, the ice melts and the water runs out of the rock. The crack is now bigger.

3

Rocks with large cracks can break apart. It took many years for this rock to break.

Read a Diagram

How can ice change rocks?

 Science in Motion Watch how rocks change at **www.macmillanmh.com**

How does Earth change quickly?

Some changes on Earth happen quickly. Fast changes on Earth can be caused by changes in Earth's crust.

◀ An **earthquake** happens when Earth's crust shakes. Earthquakes cause damage and can change the shape of the land.

▶ When a **volcano** erupts, it is an opening in Earth's mantle and crust. A very hot, thick liquid comes out of the opening. The liquid cools into solid rock.

Other fast changes on Earth are caused by water. A **flood** can happen when a lot of rain falls quickly. The rain can not soak into the ground. Floods make it hard for plants and animals to live.

▲ A landslide happens when the rocks and soil slide from higher to lower ground. Rain or melting snow can cause landslides.

✓ How can a volcano change Earth quickly?

Think, Talk, and Write

1. **Predict.** How might Earth's surface change after a flood?

2. How can water change Earth slowly?

3. Draw and write about one way the surface of Earth can change quickly.

Art Link

Draw pictures to show one way the land around you has changed.

LOG ON ⓔ-Review Summaries and quizzes online at www.macmillanmh.com

Living with Floods

▲ **People build their houses on stilts to keep them safe from floods.**

In some places floods happen each year during certain seasons. People in these places have learned to put the floods to use.

Wet Season

Each year in Vietnam, the wet monsoon season lasts from May to November. Floods cover flat fields where the farmers grow their rice. Rice needs a lot of water to grow.

rice farm

Dry Season

The dry season starts in November. The Mekong river dries up and salty water flows in from the sea.

Rice can not grow in the salty water. Some farmers build barriers to keep the seawater from entering. Other farmers turn their rice fields, or paddies, into shrimp farms.

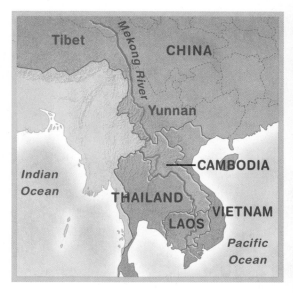

The Mekong is one of the longest rivers in the world. It starts in Tibet and ends in Vietnam.

shrimp farm

Talk About It

Predict. What will farmers turn their shrimp farms into in May? Why?

AMERICAN
MUSEUM of
NATURAL
HISTORY

What on Earth!

Earth's land can be flat or rocky. Earth's land can be high or low. What is the land like where you live?

Salt water fills Earth's oceans. Fresh water fills the rivers, lakes, and ponds. It also falls as rain or snow. How did you use Earth's water today?

Blowing sand and moving water can change the shape of Earth's land. Freezing water and melting ice can cause rocks to break. This is how Earth changes slowly.

Earthquakes, floods, and landslides cause damage. When lots of rain falls at once, rivers can flood. This is how Earth changes quickly.

CHAPTER 5 Review

Vocabulary

Use each word once for items 1–6.

core
crust
earthquake
landform
landslide
mantle
ocean
volcano

1. Earth's surface can change quickly because of a _____, _____ or _____.

2. A large deep body of salty water is called an _____.

3. Each different shape of Earth's land is called a _____.

Use this diagram to answer items 4–6.

4. This is Earth's _____.

5. This is Earth's _____.

6. This is Earth's _____.

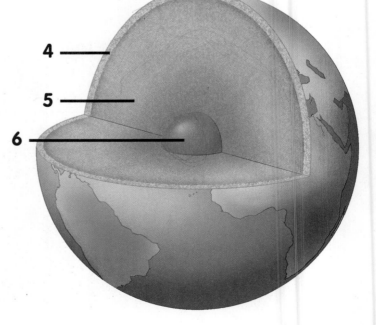

Answer the questions below.

7. **Make a Model.** What does this
model show?

8. **Compare and Contrast.** How is a hill
different from a mountain? How are
they alike?

9. Describe the different kinds of land
on Earth.

10. How can we describe Earth's land and water?

CHAPTER 6

Earth's Resources

The **Big Idea**

How do we use Earth's resources?

Burnley Falls, California

Key Vocabulary

natural resource materials from Earth that people use in daily life (page 188)

minerals bits of rocks and soil that plants and animals need (page 190)

soil a mix of tiny rocks and bits of dead plants and animals (page 196)

pollution anything that makes air, land or water dirty (page 204)

Rocks and Minerals

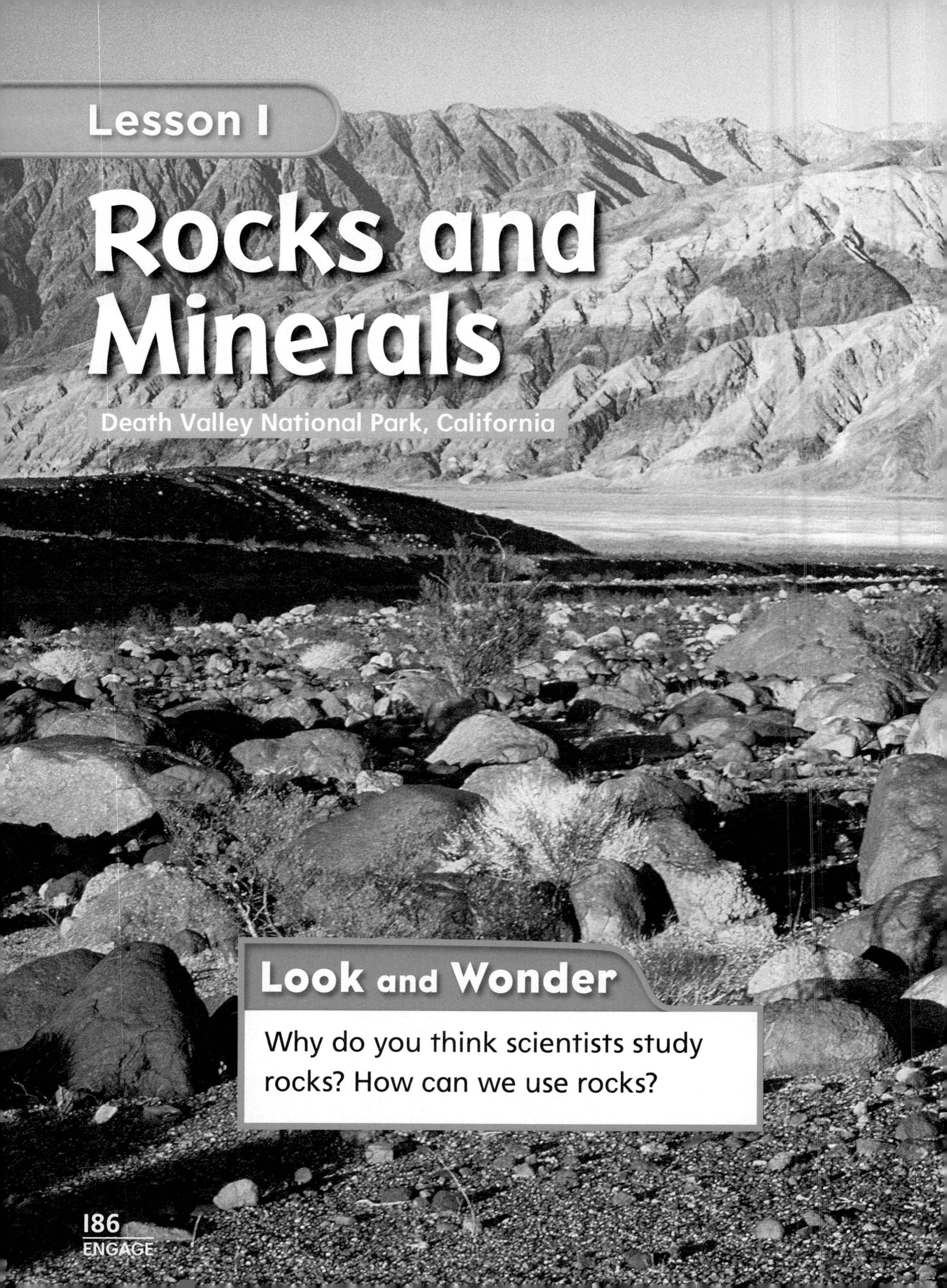

Death Valley National Park, California

Look and Wonder

Why do you think scientists study rocks? How can we use rocks?

How can we sort rocks?

What to Do

1. **Observe.** Look at your rocks with a hand lens. Describe what you see. How are they alike? How are they different?

2. **Classify.** Put your rocks into groups. Write your groups on a chart. Record how many rocks are in each group.

3. **Communicate.** Share your chart with a partner. Discuss how you put the rocks into groups.

Explore More

4. What other ways can you classify rocks?

You need

rocks

hand lens

Step 2

Read Together and Learn

Vocabulary

natural resource

rock

minerals

SCIENCE QUEST Explore rocks and minerals with the Treasure Hunters.

What are rocks?

Things people use from Earth are called **natural resources**. We use natural resources every day. Some of Earth's resources are air, water, plants, animals, and rocks.

A **rock** is a nonliving part of Earth. Most rocks are hard. Rocks can be different shapes and sizes. They can have a different feel or color.

◄ Rocks come in all shapes and sizes.

FACT Some rocks are softer than your fingernail!

Rocks cover Earth. They are found below city streets. They are found below grass and soil. Rocks are even found at the bottom of the ocean!

How do we use rocks as resources? Rocks have been used as tools for thousands of years. Many rocks can be carved, chipped, or ground.

▲ Long ago, people ground this ax head from rock.

▲ This statue in Egypt, called the Sphinx, was carved from rock thousands of years ago. It is made from one huge piece of limestone.

 How do people use rocks?

What are minerals?

Have you ever looked at a rock and seen it sparkle? Minerals in rocks can make them shine. **Minerals** are hard, nonliving parts of soil. Rocks can be made of one or more minerals. The chart shows some ways people use minerals every day.

This rock is called granite. It is made of the minerals mica, quartz, and feldspar.

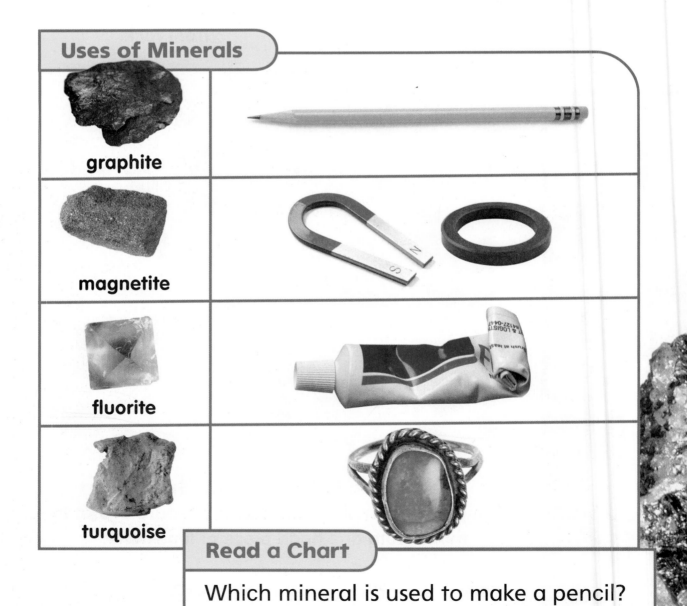

Uses of Minerals

graphite	
magnetite	
fluorite	
turquoise	

Read a Chart

Which mineral is used to make a pencil?

Rocks and minerals form in the ground over millions of years. People dig in the ground to find minerals.

 How are rocks and minerals different?

Quick Lab

Observe minerals with a hand lens. List what makes each mineral different.

A rock hammer is used to break rocks.

Think, Talk, and Write

1. **Classify.** Choose four rocks. Classify them by their shape, size, color, and feel.

2. Where can you find rocks and minerals?

3. Write about some ways that rocks are alike and different.

Math Link

Find three rocks where you live. Put the rocks in order by size. Then measure the rocks to see if your order was correct.

LOG ON **e-Review** Summaries and quizzes online at **www.macmillanmh.com**

Rock and Stroll

Rocks are all around us. We see them at the beach and in the mountains. We see them in gardens and on playgrounds. Go for a walk and look at rocks. Take notes about the rocks you see.

✏ Write About It

Write a letter to a friend. Write about your walk. Describe the rocks you saw. Explain how you think they got their shape.

Remember
A letter shares news or ideas with someone.

LOG ON ⓮ –Journal Write about it online at **www.macmillanmh.com**

Rock Patterns

You can use rocks to make patterns. Look at the pattern below. What kind of rock do you think would come next? How do you know?

Make a Pattern

Use rocks or draw pictures of rocks to make a pattern. Share it with a partner. Have your partner explain which rock they think will come next.

Remember
A pattern has a unit that repeats.

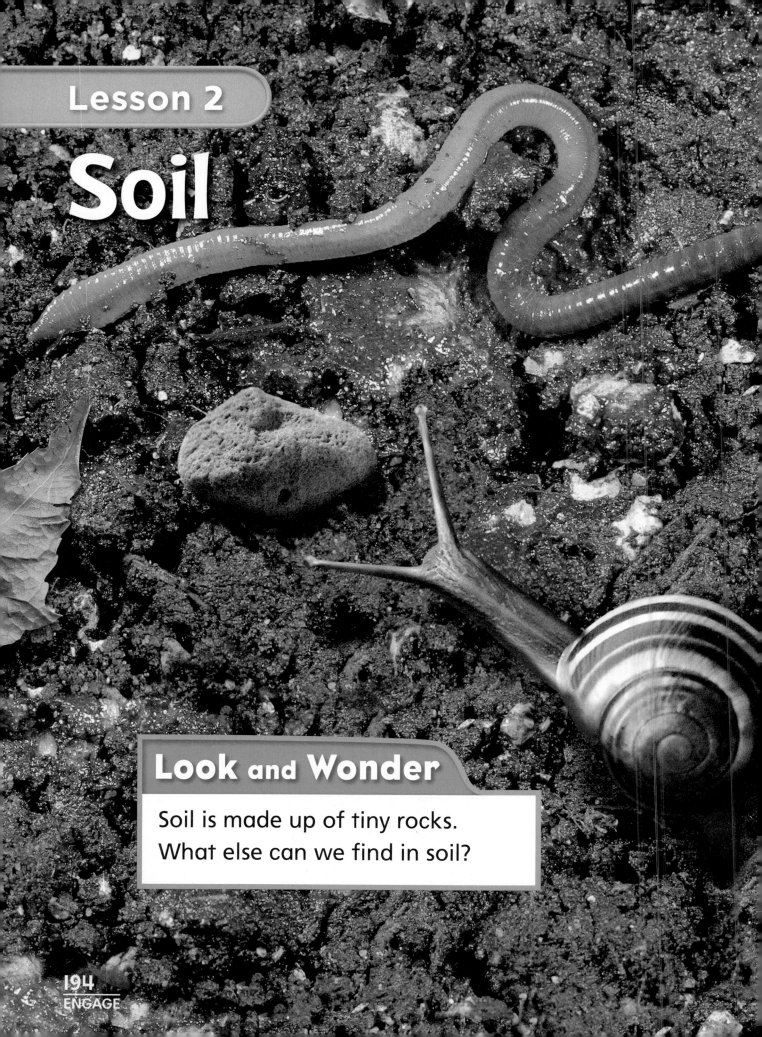

Soil

Look and Wonder

Soil is made up of tiny rocks.
What else can we find in soil?

What is in soil?

What to Do

1. Put some soil in a strainer. Gently shake it over a plate.

2. **Observe.** Look at the soil on the plate. Use a hand lens. Draw what you see.

3. Pour the soil left in the strainer onto another plate. Observe the soil. Draw what you see.

Explore More

4. **Communicate.** Use some new soil. Repeat this activity.

Step 1

You need

soil

2 plates

strainer

hand lens

What is soil?

Soil is a mix of tiny rocks and bits of dead plants and animals. These small pieces become part of the soil and help plants grow.

Soil can have different colors and textures. It can have bits of rock of different sizes.

Types of Soil

clay soil

topsoil

Some soils are darker and hold more water. Other soils are very rough, like gravel. Some soils are sandy and others feel more like clay. Sometimes there are tiny pieces of broken rock called silt in soil. Silt and clay pieces are so small they feel smooth between your fingers.

✓ **What makes up soil?**

Read a Photo

Describe each soil.

sandy soil

How is soil formed?

Soil takes a very long time to form. Over time, rocks and minerals break down into smaller pieces. Plant and animal parts rot, or **decompose**.

Mushrooms help break down the dead plants. Nutrients that were once in living things become part of the soil. These nutrients make the soil healthy.

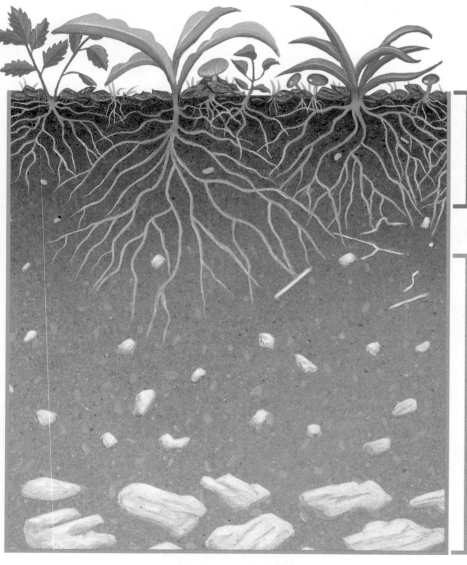

◄ This layer of soil is topsoil. It is best for growing plants. It has mushrooms and rotting plant and animal parts.

◄ This layer of soil beneath topsoil is called subsoil.

You can see things decompose in the soil by making a compost pile. Compost is a mix of soil and rotting parts of plants and animals.

 Which layer of soil is best for growing plants?

◀ **This rotting log will decompose and become part of the soil.**

Think, Talk, and Write

1. **Sequence.** How is soil formed?

2. What are some of the differences between soils?

3. Write about ways people use soil.

Health Link

Make a list of plants that grow in soil. Place a check next to the plants that we use for food.

LOG ON **e-Review** Summaries and quizzes online at **www.macmillanmh.com**

199
EVALUATE

Using Earth's Resources

Look and Wonder

What natural resources do you see in this picture? How are people caring for them?

How do we use Earth's resources every day?

What to Do

1 Make a chart about how you use water, air, plants, animals, and rocks.

2 **Communicate.** Write down your ideas on the chart.

3 Work with a partner. Think of other things you use from Earth. Write down your ideas.

Step 3

4 **Draw Conclusions.** How are the things that come from Earth important to us?

Explore More

5 **Infer.** What if there were no more water or rocks on Earth? How would your life change? Write your ideas.

Vocabulary

pollution

reduce

reuse

recycle

How do we use natural resources?

Air, wind, water, rocks, and soil are natural resources we use every day. Natural resources give us many of the things we need to live. Some resources, such as water and wind, are replaced quickly by Earth.

▲ Moving water can make power to light and heat homes.

▲ Trees can be used for building homes and furniture.

Wind can also make power to light and heat homes.

Some natural resources are not made quickly by Earth. It can take millions of years for minerals to form. We use minerals every day. Once these resources are gone, they can not be quickly replaced.

 Which resources can be replaced quickly?

Animal parts can be used to make clothing and other products.

Soil is used to grow crops. Soil is not made quickly by Earth.

▲ **Oil comes from deep in the ground and is used for fuel.**

Coal is a rock that is taken from the ground. It can be used to heat homes. ▼

Why should we care for Earth's resources?

We need Earth's land, air, and water to live. When these resources are wasted or made dirty, we can not use them.

Pollution is anything that makes water, air, or land dirty. Factories and cars can make air pollution. Oil spills and other garbage in the water can hurt animals.

▲ **Cut up drink can rings before you throw them out, so animals will not get caught in them!**

Pollution

FACT ▶ Drinking water is not quickly replaced. It has to filter through the ground for a long time.

Litter is garbage that people leave behind. People can pick up litter to help stop land and water pollution. After having a picnic or playing outside, it is important to clean up garbage.

 Why is it important to keep Earth's resources clean?

These children are cleaning up litter.

Read a Photo

How might pollution hurt the living things in this river?

LOG ON *Science in Motion* See how pollution dirties Earth's water at **www.macmillanmh.com**

How can we save Earth's resources?

When we conserve something, we save it for use in the future. We can help to conserve Earth's resources. There are three R's you can remember to help save resources. They are Reduce, Reuse, and Recycle.

◀ **Reduce means to cut back on how much you use something. Shut off the water while you brush your teeth!**

▲ **Reuse means to use something again. This egg carton now holds paint!**

Recycle means to make new items out of old items. Paper, glass, plastic, and metals can be recycled. ▶

We can also make less pollution in the air and on land. How can we help? Use bicycles more and drive in cars less. Put trash in garbage cans.

 Why is it important to reduce, reuse, and recycle?

Quick Lab

Set up bins for reusing and recycling in your classroom. **Predict** how much you will save in one week.

Think, Talk, and Write

1. **Problem and Solution.** What harms Earth's resources? How can we help Earth?

2. What are some of Earth's natural resources?

3. Write about how humans can pollute Earth.

Art Link

Make a poster showing a natural resource. Write on the poster why we should care for that resource.

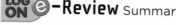 **e-Review** Summaries and quizzes online at www.macmillanmh.com

A World of Wool

Where does wool come from? Many people all over the world use sheep or goats to get wool. Some people get wool from other animals. Scientists at the American Museum of Natural History collect letters from people around the world. The scientists learn how people in other countries get wool.

Peru is a country in South America.

Dear Museum,

My name is Juana. I live in the Andes Mountains of Peru and the weather is cold. My family and I wear sweaters to keep us warm. Our sweaters are special because they are not made of sheep's wool. They are made of wool from llamas.

Llamas look like small camels. They have long necks and long legs. They have thick fur to keep them warm in the mountains. Some farmers in Peru raise llamas for their fur.

My wool sweater keeps me warm and dry in the winter!

Bye for now,

Juana

▲ Workers spin the wool from the llama into yarn.

Talk About It

Problem and Solution. What wool do some people use where there are no sheep?

What Earth Gives Us

Soil is a useful resource. Tiny bits of some plants and animals become part of soil when they die. New plants grow from the soil. We use plants for food, clothing, and building materials.

You can find rocks everywhere. Rocks can be on the tops of mountains or in the soil below your feet. Rocks are made of minerals. The minerals in soil help plants grow.

Nature gives us many resources. The Sun, wind, water, rocks, plants, and animals are just some of them. We can help protect Earth by using only what we need.

Don't litter! If you can not find a way to reuse something, try to recycle it. **Reduce, Reuse, Recycle!** You can help make this world a better place to live.

Vocabulary

Use each word once for items 1–4.

| decompose |
| natural resources |
| recycle |
| reduce |
| reuse |
| rocks |
| soil |

1. Plants, water, rocks, and minerals are examples of Earth's _____.

2. When plants and animals die, their bodies _____.

3. You can help to save Earth's resources. You can _____, _____ and _____ items that you might throw in the garbage.

4. It can take Earth thousands of years to make _____ and _____.

Answer the questions below.

5. Compare. How are these rocks alike and different?

6. How are these soils alike and different?

7. Problem and Solution. List two or more natural resources. How do we harm them? Why is it important to take care of them?

8. How do we use Earth's resources?

Gemologist

Have you seen people wearing rings or earrings with colorful stones in them? Many of these stones are called gems. Gemologists are scientists who study gems.

Gems are hard and rare minerals that are known for their beauty. Sapphires, emeralds, and diamonds are some gems.

The job of a gemologist is to identify a gem and to find out its quality and value. Gemologists use special tools to identify gems and see if they have cracks in them.

gemologist

More Careers to Think About

jewelry designer

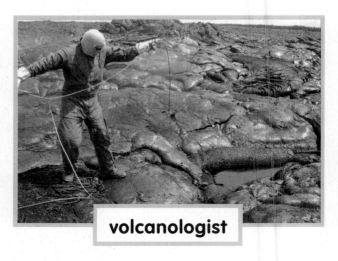

volcanologist

Weather and Sky

One million cloud droplets
make one raindrop.

Sunflakes

What if sun rays fell like snowflakes,
Drifting slowly from the sky?
So brightly they would sparkle,
Just like the Sun that's way up high.
Maybe there would be a sun day,
When we would be off from school,

When we could make sun angels
And go sunboarding, wouldn't that be cool?
I wonder what it would be like,
To catch sunflakes on my tongue.
I'm sure that sledding in the summer
Would be a lot of fun!

Talk About It

What do you think
sunflakes would look like?

CHAPTER 7

Observing Weather

The Big Idea How can we describe weather?

Key Vocabulary

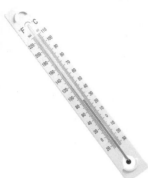

temperature
a measurement of how hot or cold something is
(page 224)

precipitation water falling from the sky as rain, snow, or hail
(page 225)

cumulus large, white, puffy clouds
(page 238)

stratus thin clouds that form into layers like sheets (page 239)

221

Weather

Look and Wonder

Weather changes from day to day. How can you describe this kind of weather?

How does the weather change each day?

What to Do

1 Make a chart with these columns at the top: Date, Temperature, Weather.

2 **Record Data.** Observe the weather each day. Record on the chart what you see. Draw any clouds you see.

3 **Compare.** After several days, compare how the weather changed from day to day.

Explore More

4 Add a column to your chart called Wind. Record how wind changes from day to day.

You need

thermometer

construction paper

Step 2

Vocabulary

temperature

precipitation

anemometer

What is weather?

When you get dressed in the morning, you might think about the weather. How hot or cold will it be outside? Is it a sunny day, or is it raining or snowing?

One way that people describe weather is by the temperature. **Temperature** describes how hot or cold something is.

This thermometer shows 70 degrees Fahrenheit (70°F) or 21 degrees Celsius (21°C).

This thermometer shows 40°F or 5°C. It is cooler, so you must dress warmly.

Another way people describe weather is by telling if it is raining. Rain, snow, sleet, and hail are all kinds of precipitation. **Precipitation** is water that falls to Earth from clouds. Each kind of precipitation can be measured in a different way.

 How can you talk about weather?

Hail is chunks of ice that fall from thunderclouds.

A ruler is used to measure the depth of snow.

A rain gauge is used to measure the amount of rain that falls in one spot.

What is wind?

All around Earth there is hot and cold air. Differences between hot and cold air cause the air to move, making wind.

Wind can be strong or light. One way to measure how the wind is blowing is with a wind sock. Wind blows into the open end of a wind sock and makes it point in the direction the wind is blowing.

If the wind sock moves gently, the wind is light. If it blows out straight, the wind is strong. ▶

Quick Lab

Make a wind sock. **Observe** how strong the wind is.

Wind Sock

Read a Photo

How strong is the wind in this photograph?

People also measure the wind's speed with an **anemometer**. The cups on the anemometer catch the wind and spin. The stronger the wind is, the faster the cups spin. The anemometer keeps track of how many times the cups spin.

This scientist is using an anemometer. ▶

 How can we measure wind?

Think, Talk, and Write

1. **Summarize.** How do anemometers and wind socks measure wind?

2. What are some different kinds of weather?

3. Write about and draw the different kinds of precipitation.

Social Studies Link

Research what the weather is like in another country. How is it like your weather? How is it different?

 e-Review Summaries and quizzes online at **www.macmillanmh.com**

A Snowy Day

What are these people doing on this snowy day?

✏ Write About It

Write a story about what you might do on a snowy day. Use the photograph to help you think about what a snowy day might be like.

Remember

A story has a beginning, middle, and end. A story uses describing words.

 e-Journal Write about it online at **www.macmillanmh.com**

A Sunny Day

Look carefully at the thermometer. What is the temperature in degrees Fahrenheit (°F)?

Write a Number Sentence

The weatherman predicted that the temperature will rise to 55 degrees Fahrenheit (55°F). Write a number sentence to show how many degrees Fahrenheit it will rise.

Remember

The left side of the thermometer shows Fahrenheit.

The Water Cycle

Torres del Paine National Park, Chile

Look and Wonder

Where do you see water in this picture?

Where did the water go?

You need

What to Do

① Fill both cups halfway with water. Mark the water levels.

② Cover one cup with plastic wrap. Tape it to the cup. Place both cups in a sunny place.

③ **Predict.** How will the levels of water change in each cup over several days?

④ **Record Data.** Write what you see in each cup every day.

⑤ **Draw Conclusions.** What happened to the water levels after several days? Why?

Explore More

⑥ What would happen if you used twice as much water? Try it.

2 cups

PLASTIC WRAP

plastic wrap

water

tape

marker

Step ②

How does water disappear?

When you jump into a swimming pool you feel the water on your body. This water is liquid.

But water is not always liquid. When water heats up, some of it changes into a gas we cannot see. Water can **evaporate**, or change from a liquid to a gas. The gas travels in the air as water vapor.

Liquid water from the pond evaporates into water vapor in the air.

FACT Clouds form everywhere, not just over water.

These clouds are made of droplets of water.

water vapor

Water vapor rises in warm air. As the air cools, it can no longer hold the water vapor. So the water changes again!

This time, the water vapor will **condense**, or turn back into liquid. The water forms into tiny droplets. Sometimes these droplets form clouds.

How can water change forms?

What is the water cycle?

Water evaporates from oceans, rivers, and lakes. Then the water condenses into clouds and falls as precipitation. We call these changes the water cycle. All over Earth, water is always moving through the water cycle.

 How does water vapor move back down to Earth?

Quick Lab

Make a model of the water cycle. Observe how hot water changes in a closed container.

The Water Cycle

Read a Diagram

How does water get in clouds?

LOG ON *Science in Motion* Watch the water cycle at **www.macmillanmh.com**

Water is warmed by the Sun, evaporates, and rises into the air.

Once the air cools, the water condenses and forms clouds.

Think, Talk, and Write

1. **Cause and Effect.** What happens in the water cycle?

2. What happens when water evaporates?

3. Write about how water changes when it evaporates and condenses.

Music Link

Write a song about water vapor and clouds.

LOG ON e-Review Summaries and quizzes online at **www.macmillanmh.com**

Precipitation can fall as rain, snow, sleet, or hail.

Over time, rain and melted snow flow back to the lakes, rivers, and oceans.

Changes in Weather

Owens Valley, California

Look and Wonder

Look at the sky. Did you know that clouds can help predict weather?

How can clouds help predict the weather?

What to Do

1 **Observe.** Look carefully at the sky every day this week.

2 **Record Data.** Draw the kinds of clouds you see each day for a week. Write the date next to each picture. Then predict what the weather will be like tomorrow.

3 The next day, record what the weather is like. Draw the clouds and the date. Was your prediction from the day before correct?

4 **Draw Conclusions.** How can clouds help predict weather?

mostly blue sky

puffy clouds with dark purple edges.

our house

Tuesday, March 10.

I predict _it will rain._

What happened: _I was right. It rained!_

Explore More

5 **Predict.** Write a weather report for the next week. Why is tomorrow's weather the easiest day to predict?

What are different kinds of clouds?

Not all clouds look the same. There are different kinds of clouds. Each means a different type of weather may be coming.

Cumulus clouds are small, white puffs. Cumulus clouds may appear in long, rippled rows.

▼ Cumulus clouds are often seen in spring and summer. Small, puffy cumulus clouds mean good weather.

FACT ▷ Not all clouds are rain clouds.

▲ When you see cirrus clouds, it usually means that the weather will change within the next day.

Cirrus clouds are thin clouds very high in the sky. They are made of ice. The wind blows cirrus clouds into wispy streams.

Stratus clouds are often low in the sky. They come in sheets and cover the entire sky. Stratus clouds can be thick or thin.

 Which cloud type is best for the day of a picnic?

▲ When you see stratus clouds, it usually means that a storm with rain or snow is coming.

How can we stay safe from weather?

When different types of air come together, the weather changes. Storm clouds can grow thick and lightning can form in them.

Stay safe in dangerous weather. Pay attention to weather reports. During storms, do not go outside near tall objects where lightning may strike.

Quick Lab

Use a paper bag to make a model of a thunder sound.

Lightning Safety

Avoid open spaces.

Stay out of water.

Do not stand under trees.

Stay indoors.

Read a Chart

Where should you go in a storm?

Sometimes storms become very strong. They can turn into hurricanes and cause disasters. Strong thunderstorms can make spinning columns of air called tornadoes.

Tornadoes can pick up large objects from the ground and damage property.

 How can weather change?

Think, Talk, and Write

1. **Main Idea and Details.** How can clouds help people predict the weather?

2. What kinds of clouds predict a storm is coming?

3. Draw and write about a storm you have seen.

Social Studies Link

Make a short skit or commercial showing how to prepare for severe weather.

LOG ON e-**Review** Summaries and quizzes online at www.macmillanmh.com

Predicting Storms

weather balloon

Even on a sunny day, dangerous thunderstorms can happen quickly. Knowing that a storm is coming is very useful. If there is lightning, you would know to stay inside and not use electricity.

If there is a flood, you would know what to do. You can go to a higher floor in your home. How do scientists predict storms?

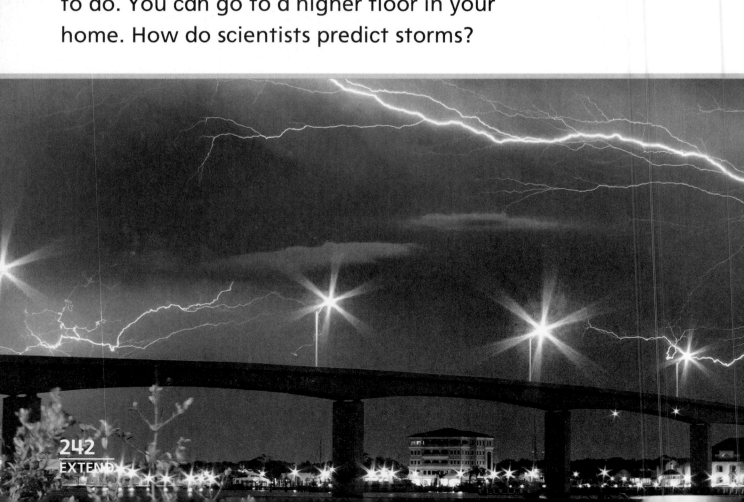

1892: Weather balloons are first used. Weather balloons can float high above Earth. As the balloons travel, they collect data about wind and weather conditions that might lead to storms.

▼ **Doppler radar**

Mid-1990's: Doppler radar is widely used. Radar is a tool that sends out radio waves into the air. The waves hit things in the air, such as raindrops. Then the waves come back. The data is recorded. Doppler radar can measure how fast a storm is moving and where the storm is going.

▲ **Doppler radar recorded this image of a hurricane.**

Talk About It

Main Idea and Details. How do scientists predict the weather?

AMERICAN MUSEUM OF NATURAL HISTORY

Earth's Water Cycle

There is a lot of water on Earth! There is water in oceans, rivers, and ponds. When water on Earth warms, some evaporates, or turns into water vapor. Water vapor rises in the air.

When water vapor cools, it
forms clouds. There are many
types of clouds. Some are thick
and some are thin. Some are high
in the sky and some are low.

When clouds get heavy,
precipitation falls to Earth.
Precipitation can be rain, hail,
sleet, or snow. There can be
storms with lightning and thunder.
Look out, a storm is coming!

The storm is over now. The rain has stopped falling. The ground is still wet, but the Sun is coming out. The Sun will warm Earth's water and start the cycle again.

CHAPTER 7 Review

Vocabulary

Use each word once for items 1–6.

condense

cumulus

evaporate

precipitation

stratus

temperature

1. When water vapor starts to _____ it turns into a liquid.

2. The _____ tells how hot or cold it is.

3. Rain and snow are kinds of _____.

4. Small, white puffy clouds are called _____ clouds.

5. You can see the Sun through these _____ clouds.

6. Water can _____. It becomes water vapor and goes into the air.

Answer the questions below.

7. Predict. Look at the photo. What kind of weather is coming?

8. Summarize. Describe the water cycle.

9. What do clouds tell us about weather?

10. What tools can we use to measure weather?

The Big Idea

11. How can we describe weather?

CHAPTER 8

Earth and Space

The Big Idea

What can we see in the night sky?

A view of Earth from the Moon

Key Vocabulary

axis a center line that an object spins around (page 255)

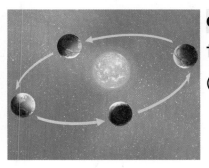

orbit the path Earth takes around the Sun (page 262)

phase the Moon's shape as we see it from Earth (page 271)

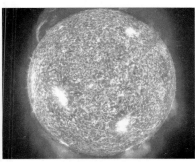

star an object in space made of hot, glowing gases (page 272)

Day and Night

Santorini, Greece

Look and Wonder

Why do you think the sky gets dark each night?

Why can't we see the Sun at night?

You need

flashlight

What to Do

① Stand 12 steps away facing a partner.

② Point a flashlight at your partner. The flashlight is the Sun. The partner is Earth.

③ **Predict.** Let your partner turn around slowly in front of the flashlight. Will he or she always be able to see the light? Try it.

④ **Infer.** How does this model show why we can not see the Sun at night?

Step ①

Explore More

⑤ **Make a Model.** What pattern is made when your partner turns around in front of the flashlight three times? Try it.

What causes day and night?

Earth spins every moment of the day and night. You do not feel it, but it is happening right now. The spinning of Earth is called **rotation**.

Earth's rotation causes day and night. When one side of Earth faces the Sun, it is day. At the same time, it is night on the other side of Earth.

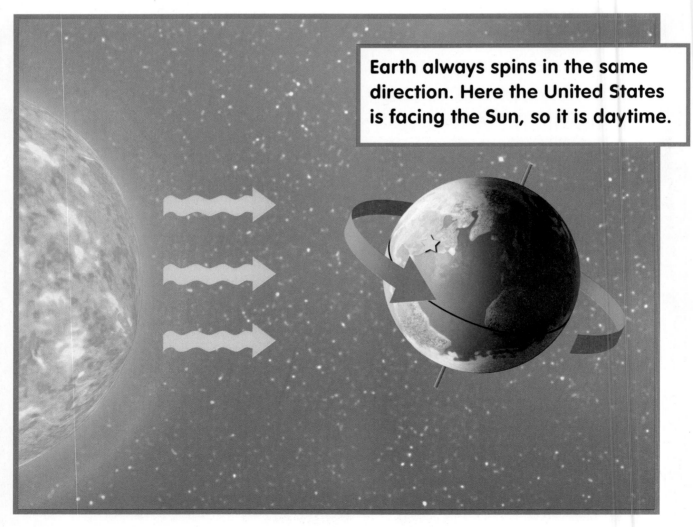

Earth always spins in the same direction. Here the United States is facing the Sun, so it is daytime.

Earth rotates around an imaginary line called an **axis**. The axis goes through the center of Earth from north to south. Every 24 hours, Earth makes one full turn on its axis. This pattern of day and night repeats again and again.

✔ Why can we see sunlight only during the day?

Read a Diagram

Is it day or night in the United States? How do you know?

Earth's Axis

axis

Why do the Sun and Moon seem to move?

We live on Earth and look out toward the sky. As Earth rotates, the Sun and Moon seem to move across the sky.

You can see the Sun make different shadows during the day. As Earth rotates, shadows on the ground change. Longer shadows mean the Sun is lower in the sky.

Quick Lab

Make a flip book of the Moon. **Observe** how the Moon seems to move across the sky in one night.

▼ The length of the shadow changes as the Sun seems to move across the sky.

8:00 a.m.

12:00 noon

4:00 p.m.

In the morning, the Sun seems to rise in the sky.

By the middle of the day, we see the Sun high in the sky.

As it gets dark, the Sun seems to set in the sky.

Each night the Moon also seems to move across the sky.

 Why does the Sun appear to move in the sky?

Think, Talk, and Write

1. **Problem and Solution.** How can you tell what time it is if you do not have a watch?

2. What causes day and night?

3. Draw and write about how the Sun or Moon seem to move.

Music Link

Write a song about day and night to the tune of "Twinkle Twinkle Little Star."

 e-Review Summaries and quizzes online at **www.macmillanmh.com**

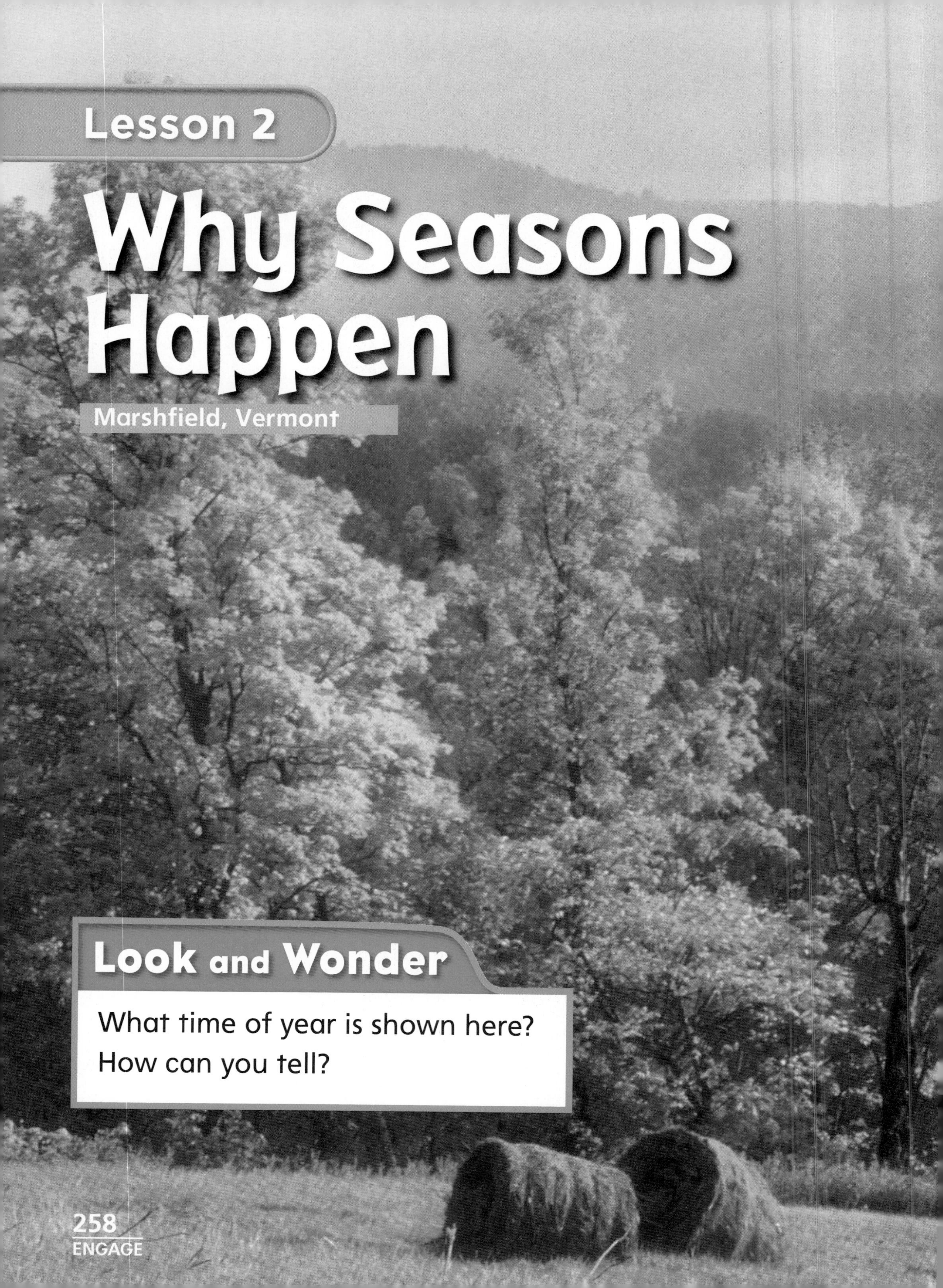

Lesson 2

Why Seasons Happen

Marshfield, Vermont

Look and Wonder

What time of year is shown here? How can you tell?

What clothes do people wear in each season?

What to Do

1 Write the name of a different season in each corner of your paper.

2 Cut out pictures of different kinds of clothes from magazines.

3 **Classify.** Glue the pictures near the seasons where they belong.

4 **Draw Conclusions.** What do people wear in different seasons?

Explore More

5 **Classify.** Sort your clothes at home by season. Explain how you grouped your clothes.

You need

paper

markers

magazines

scissors

glue stick

Step 2

What are the seasons like?

Each season has a different kind of weather. In fall, the air can become cool. Leaves on some trees turn colors and fall off.

In winter the air is cold. In some places it snows. Animals must keep warm. Some birds fly to warmer places. People wear warmer clothes.

▲ There are fewer hours of daylight in fall.

▲ In many places it can snow in winter.

In spring the weather becomes warmer. There are many rainy days. Trees and flowers bloom. Birds return from their winter homes.

Summer is the hottest season. There are more hours of sunlight than of night. What season comes after summer? The seasons start all over again!

Quick Lab

Divide a plate into four parts to show each season. Draw and **communicate** what you do in each season.

▲ **There are many rainy days in spring.**

▲ **The days are hot and long in summer.**

How is summer different from winter?

What causes the seasons?

Did you know that Earth moves around the Sun? The path Earth takes around the Sun is called its **orbit**. Earth takes about 365 days, or one year, to orbit the Sun.

We know there is day and night because Earth spins on its axis. The axis also is tilted. Earth always tilts in the same direction on its axis.

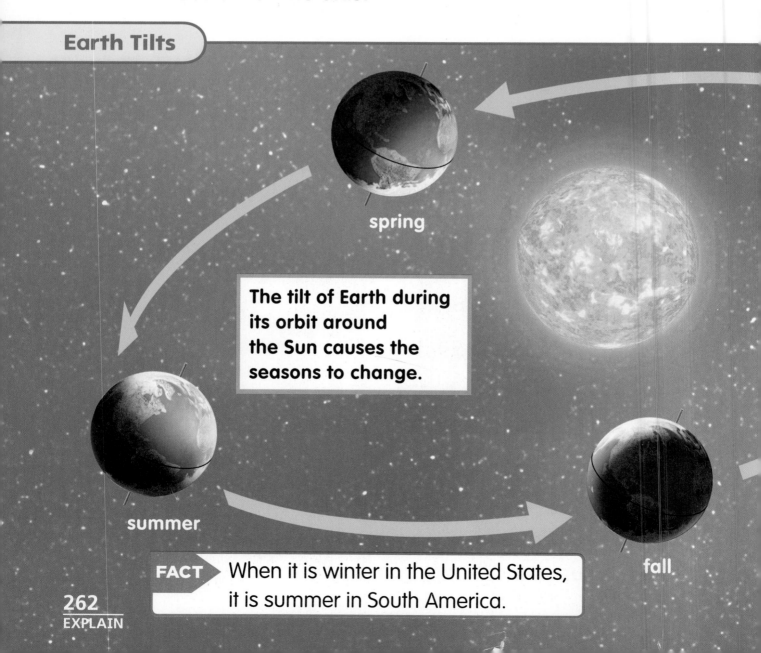

Earth Tilts

spring

The tilt of Earth during its orbit around the Sun causes the seasons to change.

summer

fall

FACT ▸ When it is winter in the United States, it is summer in South America.

As Earth moves around the Sun, the tilt of Earth causes the seasons. The part of Earth that tilts toward the Sun is warmer. The part of Earth that tilts away from the Sun is colder.

 What happens on Earth during one orbit around the Sun?

winter

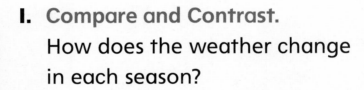
Read a Diagram

Which season shows the top half of Earth tilted away from the Sun?

 Science in Motion Watch how Earth tilts at **www.macmillanmh.com**

Fun with the Seasons

Think about the seasons and the different things you do all year. Use the photographs to help you think about what you like to do.

✏️ Write About It

Write a story to compare what you do in winter and in summer. Include details about how the seasons are alike and different.

Remember

Writing to compare tells how things are alike and different.

LOG ON ℮-Journal Write about it online at www.macmillanmh.com

How Much Sunlight?

We get more sunlight in the summer than we do in the winter. How many hours of sunlight do we get in each season? Use this chart to find out.

Hours of sunlight in one day

summer	$14\frac{1}{2}$ hours
fall	$11\frac{1}{2}$ hours
winter	10 hours
spring	12 hours

Put in Order

Put the seasons in order from least to most hours of sunlight. Make a new chart to show this.

Remember

Look at the whole numbers first to put numbers in order.

The Moon and Stars

Look and Wonder

The Moon is bright in the night sky. Where does the Moon's light come from?

How do we see the Moon at night?

What to Do

1. Use a white ball as the Moon. Turn out the room lights. Is it easy to see the Moon?

2. **Make a Model.** Shine a flashlight on the Moon. The flashlight is the Sun. Is the Moon easier to see now? Why?

3. **Draw Conclusions.** Where does the Moon's light come from?

Explore More

4. **Investigate.** What if the Moon were a different color? How would that affect the brightness of the Moon? Make a model to find out.

Step 2

Why can we see the Moon from Earth?

The Moon does not shine the way the Sun does. The Moon is made of rock! We see the Moon because light from the Sun shines on the Moon.

Look at the picture below. Point to where it is night on Earth. Then point to the part of the Moon that is lit by the Sun. You sometimes see this part of the Moon at night.

How the Moon Moves

The Sun's light shines on the Moon.

Sun

Moon

Read a Diagram

When can we see the most light on the Moon?

The Moon does not just stay still in the night sky. The Moon moves in a path around Earth. It takes the Moon about a month to make one orbit around Earth. The Moon's path around Earth repeats again and again.

 Why can we see the Moon?

▲ **The Moon has a light color because it is covered in dust.**

Earth

Why does the Moon seem to change shape?

From Earth the Moon looks as if it is changing shape. The Moon does not really change shape. Our view of the Moon changes as the Moon moves during one month.

New Moon

When the Moon is between Earth and the Sun, we can not see the Sun's light shining on the Moon. It looks as if there is no Moon at all!

First Quarter Moon

After a week the Moon looks like this. It is called the First Quarter Moon. The Moon has completed one quarter of its orbit around Earth.

◀ **The Moon is Earth's closest neighbor in space.**

On different nights we see different amounts of sunlight shining on the Moon. Each shape of the Moon we see during one month is called a **phase**. The phases appear in the same order every month. The phases repeat each month.

 What happens as the Moon orbits Earth?

Full Moon

Last Quarter Moon

The Moon moves to a new place by the next week. We can see all of the Moon's lit side. This phase is called the Full Moon.

By the third week, the Moon is three quarters of its way around Earth. It is called the Last Quarter Moon.

FACT ▶ The Moon can sometimes be seen during the day.

What are stars?

A **star** is an object in space made of hot, glowing gases. The gases give off heat and light. Some stars are very bright. Stars can be different colors and sizes.

Some stars make patterns in the sky. Stars seem to move across the sky during one night.

Quick Lab

Observe the night sky. Collect data about the stars you see. Communicate what you see to your class.

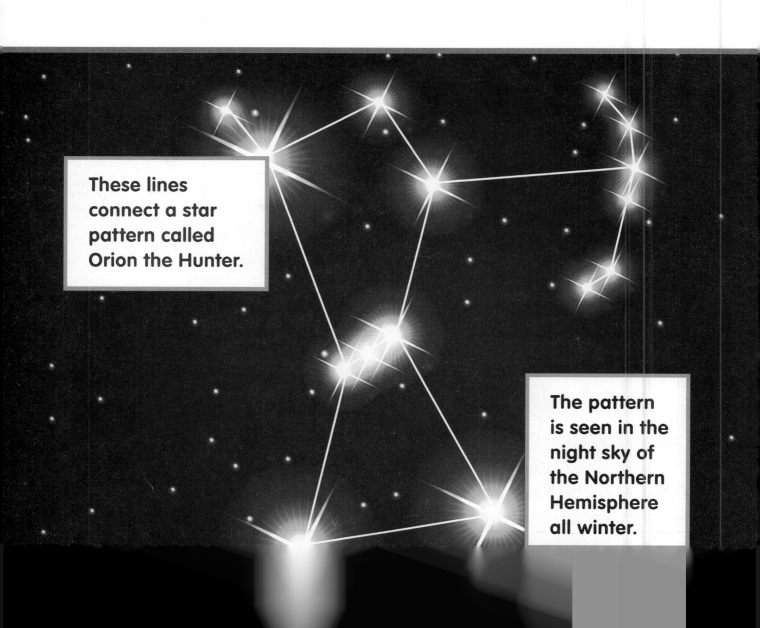

These lines connect a star pattern called Orion the Hunter.

The pattern is seen in the night sky of the Northern Hemisphere all winter.

From Earth, stars look like tiny points of light. They look tiny because they are far away.

There is one star that is close to Earth. That star is the Sun! The Sun is an average size star. It looks large to us because it is close to Earth.

 How are stars different from each other?

▲ The Sun lights up the sky during the day. We can not see other stars in the sky until night.

Think, Talk, and Write

1. **Predict.** What do you think will happen a week after a New Moon?

2. What star is closest to Earth?

3. Draw and write about how the Moon seems to change during one month.

Art Link

Draw a pattern of stars on paper. Connect the stars. Give your star pattern a name.

LOG ON ⊖-Review Summaries and quizzes online at www.macmillanmh.com

273
EVALUATE

The Solar System

Look and Wonder

What might you see if you traveled in space?

How are orbits alike and different?

What to Do

1. Draw a Sun in the middle of poster paper.

2. **Measure.** Draw an X 6 centimeters to the right of the Sun. Measure another 6 centimeters from that spot. Draw another X.

3. **Make a Model.** Draw a path around the Sun for each X. Each path shows an orbit.

4. **Draw Conclusions.** Which orbit is larger? How do you know?

Explore More

5. **Make a Model.** Continue drawing Xs until you have 8 Xs. Show which orbit is largest.

You need

poster paper

crayons

ruler

Step 2

Vocabulary

planet

solar system

What goes around the Sun?

We live on the planet Earth. A **planet** is a huge object that moves around the Sun. Our **solar system** is made of planets, moons, and the Sun. Solar means "of the Sun." The picture below shows the eight planets in our solar system.

The Solar System

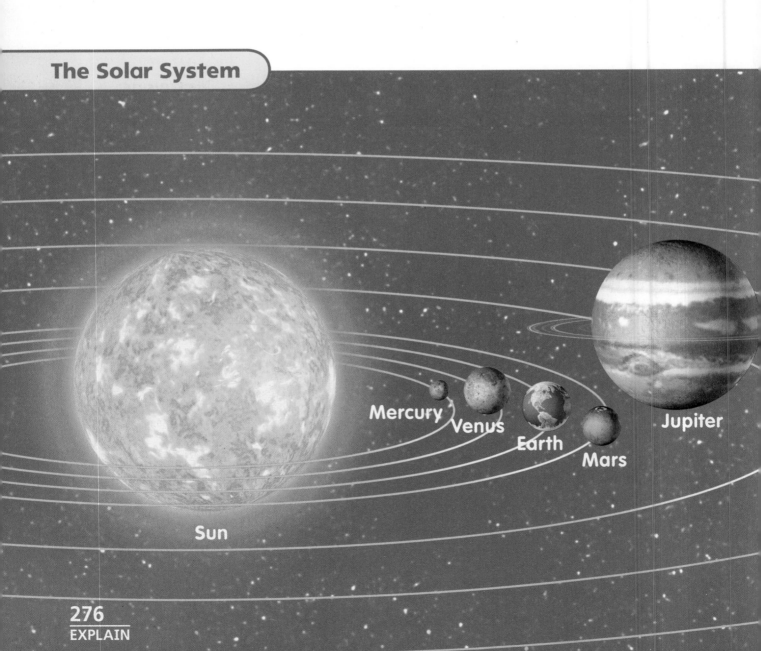

Sun

Mercury

Venus

Earth

Mars

Jupiter

The Sun is at the center of the solar system. The Sun is the strongest and the brightest part of the solar system.

Like Earth, each of the planets moves around the Sun. Planets closest to the Sun take less time to make one trip around it.

Planet	Orbit Time
Mercury	88 days
Venus	225 days
Earth	365 days
Mars	687 days
Jupiter	12 years
Saturn	29 years
Uranus	84 years
Neptune	165 years

 What is in our solar system?

Read a Diagram

How many planets are closer to the Sun than Earth is? What are they?

Saturn

Uranus

Neptune

What are the planets like?

Each planet in our solar system is different. Look at the pictures and captions to learn about each planet.

Quick Lab

Make a model of the solar system.

Inner Planets

Mercury is the closest planet to the Sun. It is rocky like our Moon.

Venus is the hottest planet! Thick clouds trap the Sun's heat.

Earth has water and air. It has one moon.

Mars has two moons. It has a red, rocky surface.

Outer Planets

Jupiter is the largest planet. It has at least 63 moons!

How are Saturn and Uranus different?

Think, Talk, and Write

1. **Sequence.** List the eight planets in order. Start with the planet closest to the Sun.

2. Which planets have rings?

3. Write about how life would be different if you lived on Mars.

Math Link

Which has the longer orbit, Saturn or Jupiter? How much longer? Use the chart on page 277.

Saturn has large rings around it. The rings are made of ice and rocks. **Saturn** has at least 47 moons.

Uranus has thin rings around it. It has at least 27 moons.

Neptune is a blue planet. It has at least 13 moons.

Starry, Starry Night

Long ago, star charts were used to guide ships.

Look at the starry night sky. How could looking at the stars help you?

Thousands of years ago, ancient sailors used the North Star to find out which way was north. This star appears in the sky closest to the North Pole. The North Star is also called Polaris. Why are these both good names for the star?

The Big Dipper and Little Dipper are each part of a group of stars. The stars form a pattern called a constellation. They can help people find their way.

Polaris, the North Star

Little Dipper

Big Dipper

Today, people still use stars for directions. Astronomers use star charts to guide telescopes in space and on Earth. The charts show millions of stars!

How to find the North Star

☆ Find the Big Dipper.

☆ Find the 2 stars that form the outside of the Big Dipper.

☆ Follow the line up to the Little Dipper.

☆ The last star on the handle of the Little Dipper is Polaris, the North Star.

Talk About It

Sequence. What first helped sailors find their way on the ocean?

AMERICAN MUSEUM OF NATURAL HISTORY

Our Moving Earth

Our planet Earth is always moving. Every day it spins. It takes all day and all night for Earth to make one full turn.

When Earth has turned part of the way around, we no longer see the Sun. Now we see the Moon and stars shining in the night sky.

spring

summer

Every day Earth also moves farther on its trip around the Sun. The whole trip around the Sun takes a year.

fall

winter

Along the way Earth has four seasons. When the year is over, the trip around the Sun begins again!

Vocabulary

Use each word once for items 1–6.

orbit

phases

planets

rotates

solar system

star

1. The planets, Moon, and Sun make up our _____.

2. The Moon's changing shapes are called _____.

3. We have day and night because Earth _____ once every 24 hours.

4. Our solar system has eight _____.

5. Every year Earth makes one _____ around the Sun.

6. This bright, hot _____ is also known as our Sun.

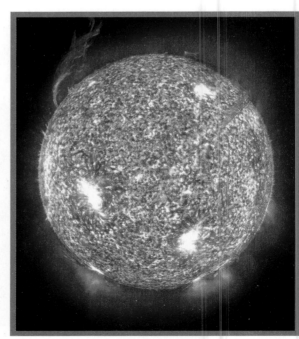

Answer the questions below.

7. What season do you think this photo shows? Why?

8. Draw Conclusions. Describe how Earth and the Moon travel around the Sun. Use balls and a flashlight to help describe what happens.

9. Compare and Contrast. What makes day and night?

10. Where does the Moon's light come from?

The Big Idea

11. What can we see in the night sky?

Careers in Science

Science Writer

If you like to read and write about science, you could become a science writer. A science writer talks to scientists about their work. Then the writer tells about the new things that scientists are learning.

Many science writers work for newspapers and magazines. They have to write about science in a way that makes it easy for others to understand.

science writer

More Careers to Think About

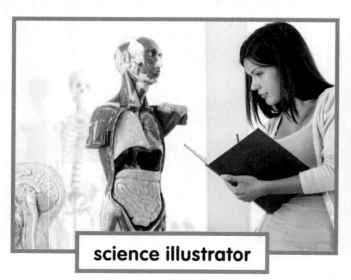

science illustrator

television reporter

LOG ON e-Careers at www.macmillanmh.com

Matter

Some paints get their color from plants and minerals.

Popcorn Hop

Round, smooth
kernels
in the pot,
Start to POP
when they get
HOT!

The soft, white
popcorn
is a treat,
All we did was
add some heat!

Talk About It
What made the popcorn
change?

Looking at Matter

The Big Idea

How can we describe matter?

Key Vocabulary

mass the amount of matter in an object

(page 296)

solid matter that has a shape of its own

(page 302)

liquid matter that takes the shape of the container it is in

(page 310)

gas matter that spreads to fill the space it is in (page 312)

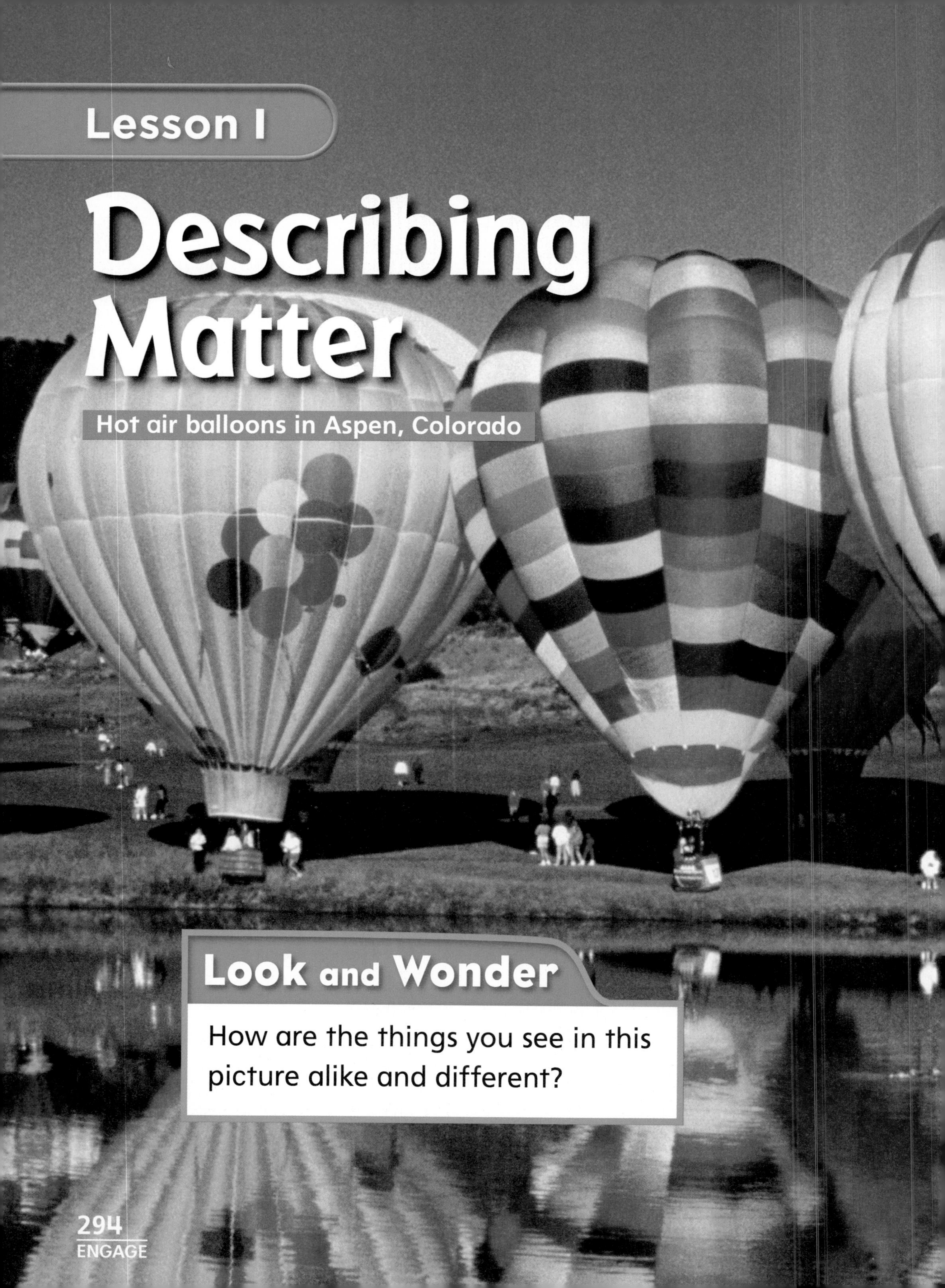

Describing Matter

Hot air balloons in Aspen, Colorado

Look and Wonder

How are the things you see in this picture alike and different?

How can you describe objects?

What to Do

You need

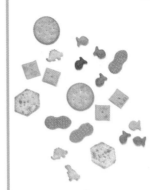

crackers

① **Observe.** Look at each cracker. Think about the different ways you can describe the crackers. What words can help you describe each one?

② **Record Data.** Make a chart like the one shown. Write your observations on your chart.

③ **Classify.** Use your chart to help you sort the crackers.

Explore More

④ How else can you sort the crackers?

	texture	shape	size	color

Vocabulary

matter

mass

property

What is matter?

Matter is anything that takes up space and has mass. **Mass** is the amount of matter in an object. The water that you drink is matter. The air you breathe is matter. Matter can be natural or made by people. We use matter every day.

Using Matter

Read a Photo

How is this boy using matter?

Different objects have different amounts of mass. A truck has a lot of mass. A pencil has a little mass. Does a book have more mass than a flower? Yes! A book feels heavier if you try to pick it up. We can use a balance to measure and compare mass.

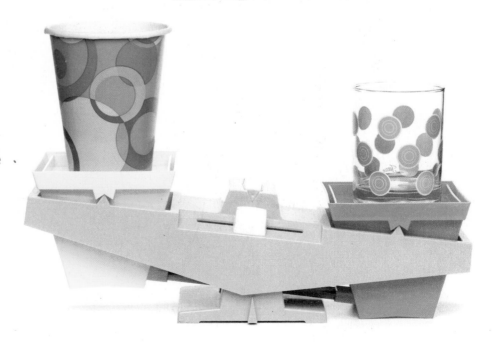

The larger shoe has more mass than the smaller shoe. ▶

Sometimes a smaller object can have more mass than a larger object. ▶

 What are some examples of matter found in your desk?

How can you describe matter?

You can describe matter by talking about its properties. A **property** is how matter looks, feels, smells, tastes, or sounds. Matter can feel smooth, rough, soft, or hard. Matter can be thick or thin. Matter can be living or nonliving.

≡Quick Lab

Classify six objects in your desk by their shape. Then sort them by their size.

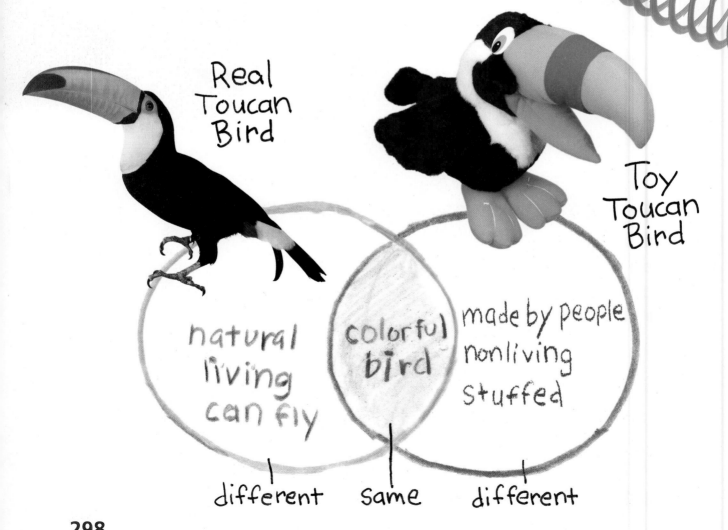

Real Toucan Bird

Toy Toucan Bird

natural living can fly

colorful bird

made by people nonliving stuffed

different   same   different

There are many ways to talk about matter. Matter can be solid, liquid, or gas.

✓ What are the properties of the things in the room around you?

◄ **This mustard is thick and gooey.**

This skunk is very smelly!

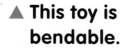

▲ **This toy is bendable.**

Think, Talk, and Write

1. **Compare and Contrast.** Choose two objects. Make a list to compare their properties.

2. What is matter?

3. Write about which has more mass, a cotton ball or a baseball.

Art Link

Use different types of matter to make a collage.

LOG ON ℮**-Review** Summaries and quizzes online at **www.macmillanmh.com**

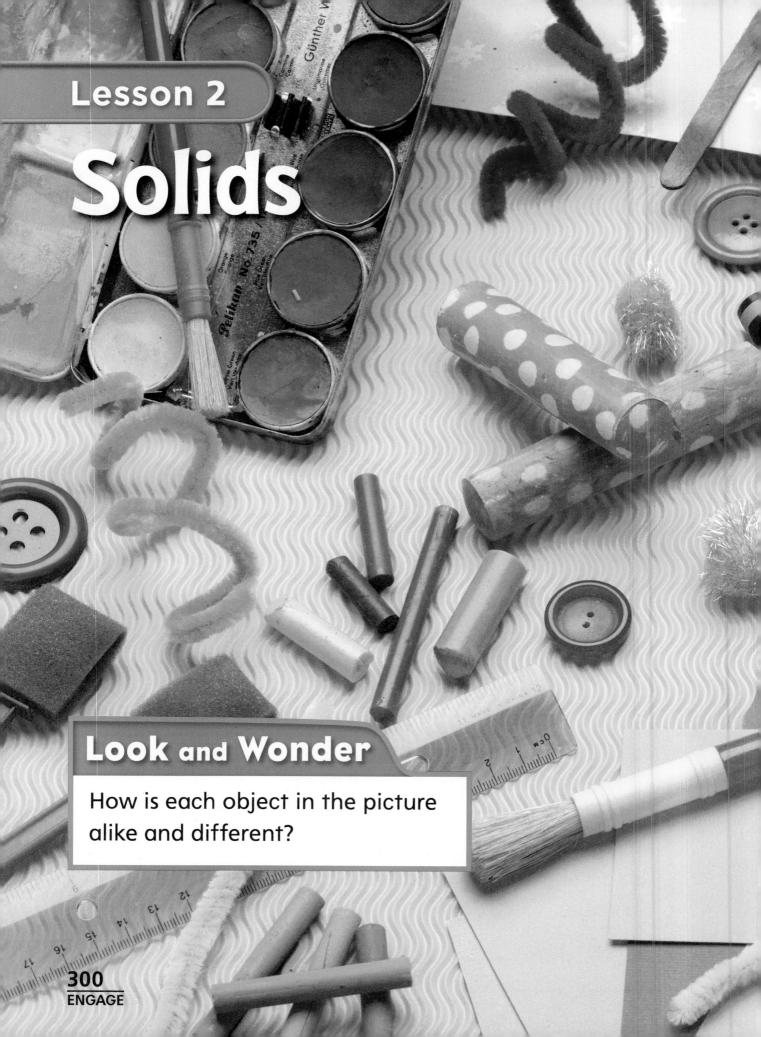

Solids

Look and Wonder

How is each object in the picture alike and different?

What are the properties of these solids?

You need

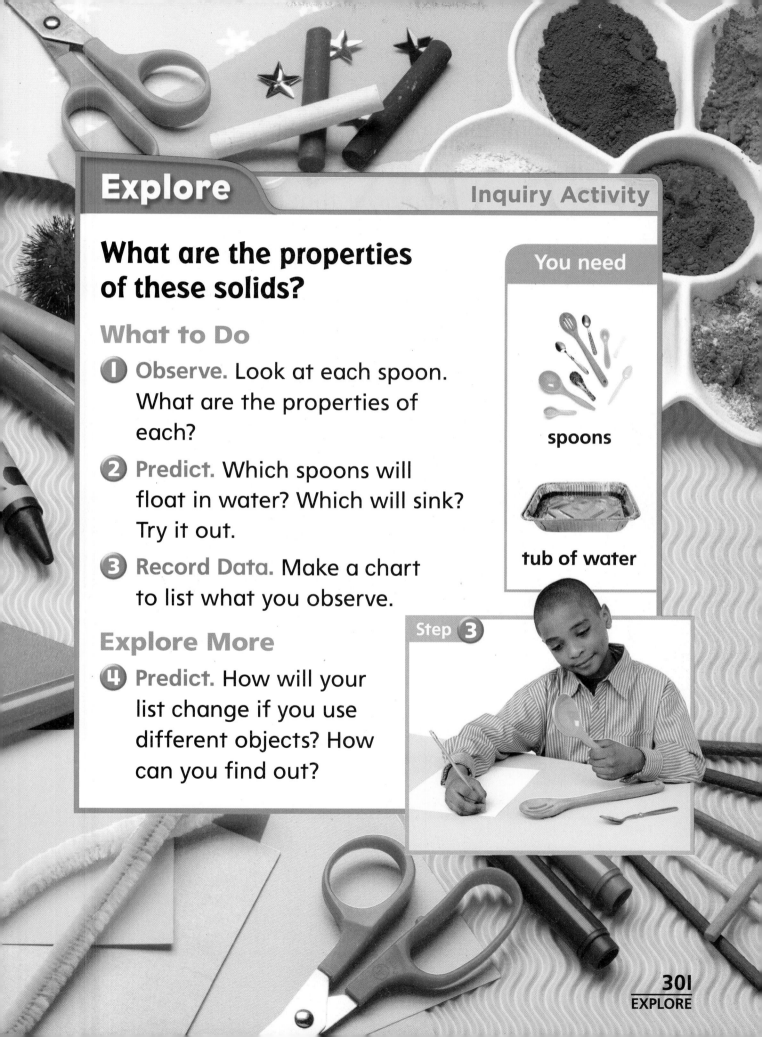

spoons

tub of water

What to Do

1. **Observe.** Look at each spoon. What are the properties of each?

2. **Predict.** Which spoons will float in water? Which will sink? Try it out.

3. **Record Data.** Make a chart to list what you observe.

Explore More

4. **Predict.** How will your list change if you use different objects? How can you find out?

Step **3**

What is a solid?

What kind of matter do you see around you? A **solid** is a kind of matter that has a shape of its own. Like all matter, solids have properties. Some solids bend. Others tear. Some solids float in water. Other solids sink.

Some Properties of Solids

rock
- hard
- speckled
- jagged

glass
- smooth
- breakable
- clear

yarn
- soft
- colorful
- long and thin

FACT ▶ Not all solids are hard.

Solids are made of different materials. Some metals, woods, and plastics are hard. Materials can be smooth or rough when you touch them. The chart below shows the properties of some solids.

 What are some properties of solids?

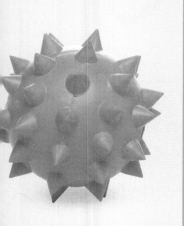

toy

- blue
- pointy
- plastic

sea sponge

- yellow
- soft
- scratchy

clay

- sticky
- bendable
- firm

How can we measure solids?

We can use tools to measure solids. A ruler tells how long, wide, or high a solid is. Some rulers measure length in a unit called a centimeter. Other rulers measure in a unit called an inch. Many rulers give both measurements.

A balance tells how much mass something has. You can measure the same object in different ways. You can measure the mass and the length of a piece of chalk.

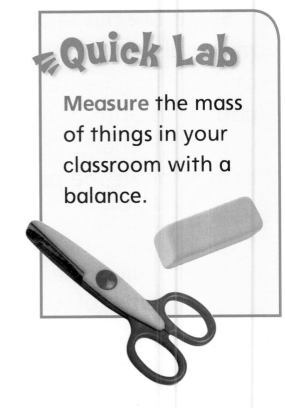

Measuring Solids

Read a Photo

What will happen to the balance if you add one more pencil to the left side?

LOG ON *Science in Motion* See how a balance measures matter at **www.macmillanmh.com**

▶ The chalk is 10 centimeters long, or about 4 inches.

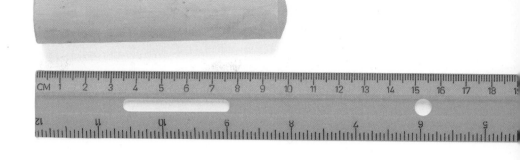

▶ Measure the distance around the chalk with string.

▶ Then measure the string with a ruler.

 What tools can we use to measure solids?

Think, Talk, and Write

1. **Summarize.** What are some properties of solids?

2. What are some examples of solid matter?

3. Write about a solid that you use every day.

Art Link

Find solids around the classroom. Make a piece of art showing some of their properties.

LOG ON e-**Review** Summaries and quizzes online at **www.macmillanmh.com**

Natural or Made by People?

This chair is made of wood. Wood is a natural product. It comes from trees. People cut down the trees. Then they shape the wood with tools to make the chair.

Wood can be painted or stained. Under the paint, the wood is still its original color.

This chair is made of plastic. Plastic is made by people. People combine chemicals to create plastic. Then they shape it in molds.

There are many different kinds of plastic. Plastic can be hard or bendable. People can also add a color to the chemicals in plastic. The plastic then becomes that color.

Which solids in your classroom are natural? Which are made by people?

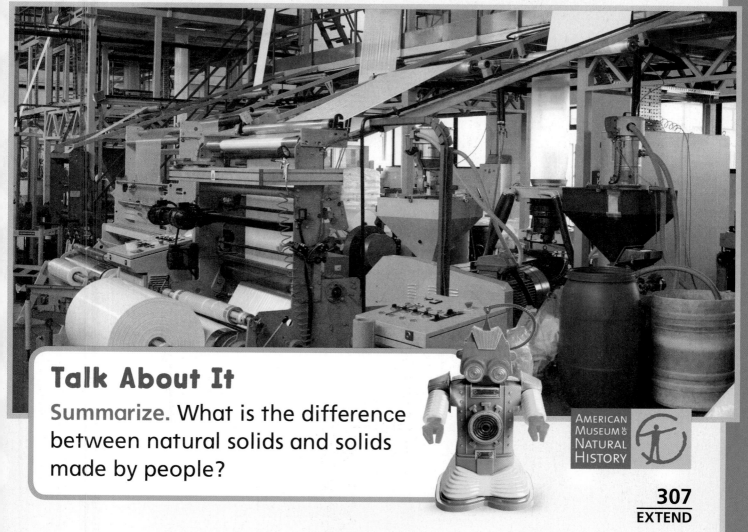

Talk About It

Summarize. What is the difference between natural solids and solids made by people?

AMERICAN MUSEUM OF NATURAL HISTORY

Liquids and Gases

Look and Wonder

Which glass is holding the most liquid?
Why do you think so?

What happens to water in different shaped containers?

What to Do

You need

1. Put containers on a tray. Measure one cup of water with the measuring cup. Pour the water into the first container. Mark where the water stops.

measuring cup

2. **Predict.** How high will the same amount of water be in the other containers?

containers

3. Pour one cup of water into the next container. Mark where the water stops. Repeat for each container.

4. **Draw Conclusions.** Were your predictions correct? Explain.

tray

Explore More

5. **Infer.** Would the activity change if you used juice instead of water? Why or why not?

Step **3**

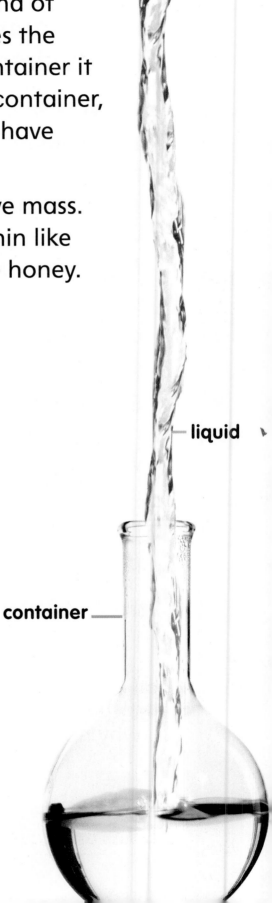

Vocabulary

liquid

volume

gas

What is a liquid?

A **liquid** is a kind of matter that takes the shape of the container it is in. Without a container, liquids flow and have no shape.

All liquids have mass. Liquids can be thin like milk or thick like honey.

liquid

container

Even in nature, liquid takes the shape of the space it is in. This waterfall flows and fills the shape of the lake.

The amount of space something takes up is called **volume**. You can measure the volume of a liquid with a measuring cup. Liquids are measured in milliliters or ounces.

The measuring cups in the picture can hold the same amount. One cup is holding a greater volume of liquid than the other.

Read a Photo

How many milliliters of liquid are in each container?

✓ What are some properties of liquids?

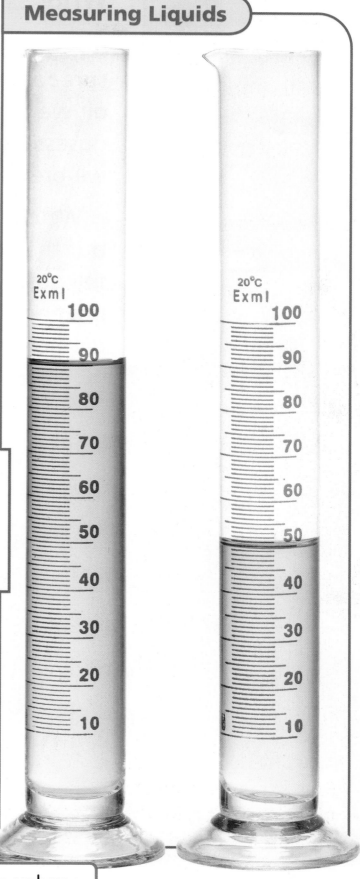

Measuring Liquids

FACT ▶ Solids and gases also have volume.

What is a gas?

A **gas** is a kind of matter that spreads to fill the space it is in. The air we breathe is made of many gases. Oxygen is one of the gases we breathe.

We can not see the gases in the air, but they are all around us. We can tell gases are there when they fill a balloon or a beach ball. We can feel air moving on a windy day.

Gases have no shape of their own.

Remember that anything that takes up space is matter. All matter has mass. How can you tell that gas has mass? Look at the picture.

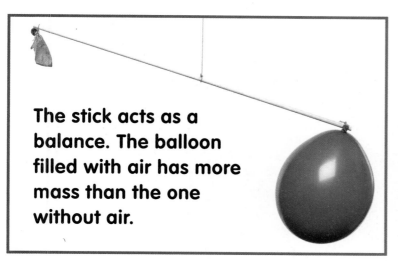

The stick acts as a balance. The balloon filled with air has more mass than the one without air.

Quick Lab

Fill containers with different kinds of matter. Have a partner **classify** the matter as solid, liquid, or gas.

 What are some properties of a gas?

Think, Talk, and Write

1. **Classify.** List the items in your refrigerator. Sort them as solid, liquid, or gas.

2. How is a gas different from a liquid?

3. Write a list of words that can be used to describe liquids. Share your list with a friend.

Health Link

Make a list of liquids that are good for you.

 e-Review Summaries and quizzes online at **www.macmillanmh.com**

Fun with Water

This girl is having fun in the water! How do you enjoy water?

✏️ Write About It

Think of times that you have had fun in water. Draw and write about what you did. Remember to add details to your story.

Remember
Details help your reader know what happened and how you felt.

LOG ON 🅔–Journal Write about it online at **www.macmillanmh.com**

Which Has More Volume?

Matt put juice in two measuring cups. What can you tell about the two containers of juice? Which has more volume?

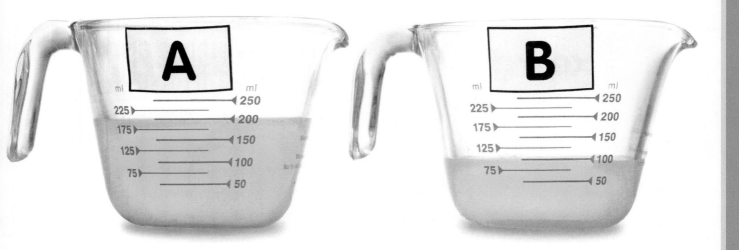

Write a Number Sentence

Cup A has 200 mL of juice. Cup B has 100 mL of juice. How many more mL are in Cup A?

Write a number sentence to show how you found the answer.

Remember

Think about which operation to choose.

Matter All Around

Matter is all around you. The clothes you wear and the water you drink are matter. Even the air you breathe is matter. Matter can be a solid, liquid, or gas.

A solid has a size and shape.
You can talk about its feel and
color. Sometimes you can talk
about its sound, smell, or taste.

Liquids take the shape of the container they are in. When a liquid spills it has no shape of its own. This girl is gliding on the flowing water.

318

Gases fill the spaces they
are in. Air fills up these rafts. The
rafts float on the water. Look at
the matter you see all around you!

Vocabulary

Use each word once for items 1–6.

| gas |
| liquid |
| mass |
| matter |
| solid |
| volume |

1. Everything that takes up space and has mass is called _____.

2. The amount of matter in something is called _____.

3. Some matter can not always be seen. It spreads to fill the space it is in and is called a _____.

4. Matter that has a shape of its own is called a _____.

5. Matter that flows and takes the shape of the container it is in is called a _____.

6. The bottle on the right can hold a larger _____ of water than the bottle on the left.

Answer the questions below.

7. **Record Data.** How are the two balls alike and different? Which ball has more mass?

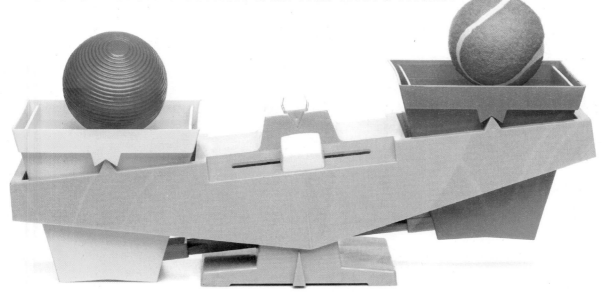

8. **Summarize.** What tools can you use to measure matter in different ways?

9. What type of matter has filled this balloon?

10. How can we describe matter?

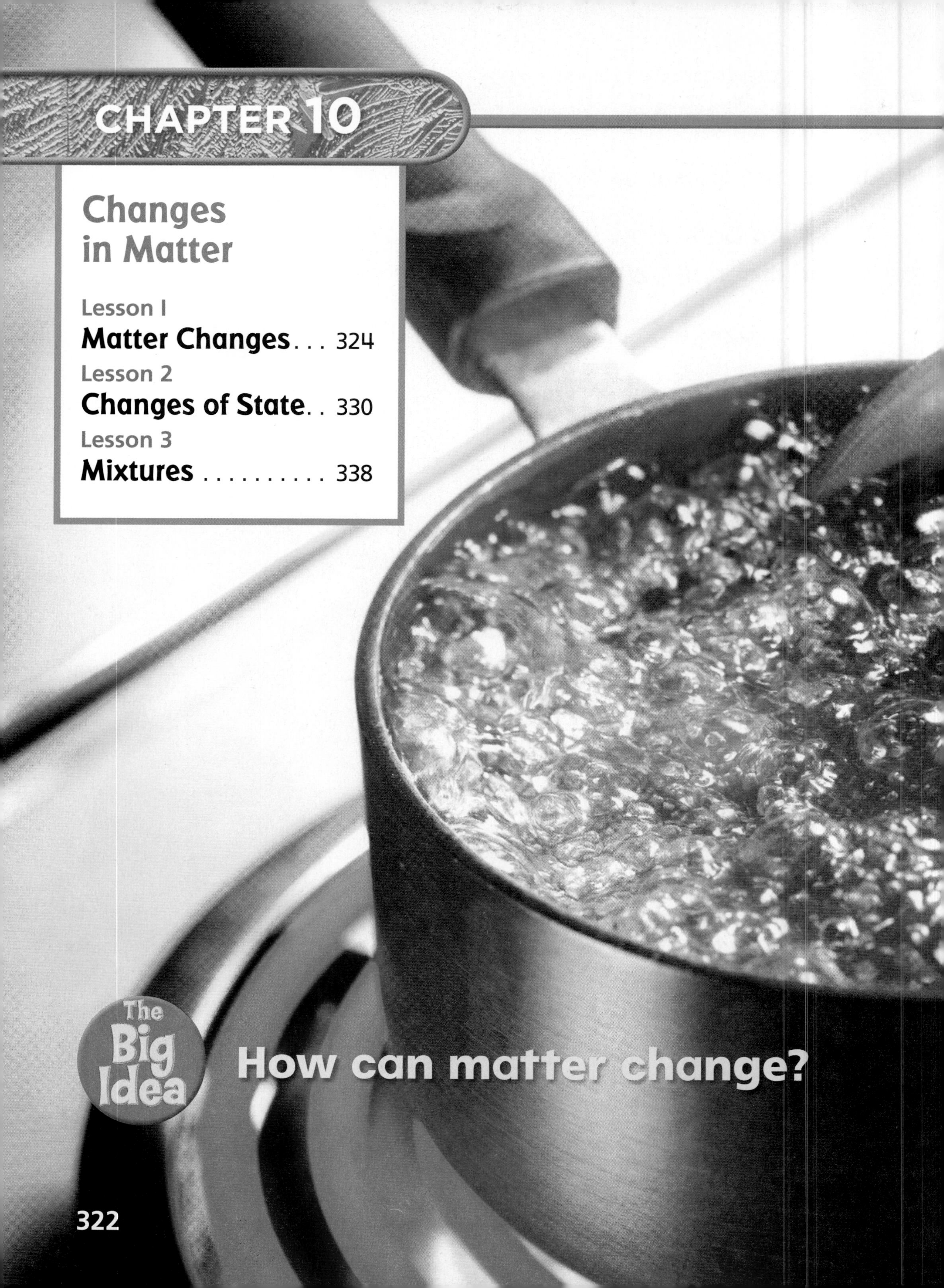

CHAPTER 10

Changes in Matter

The Big Idea

How can matter change?

Key Vocabulary

More Vocabulary

evaporate,
page 333

condense, page 334

dissolve, page 343

physical change
a change in the size
or shape of matter
(page 326)

chemical change
when matter
changes into
different matter
(page 328)

mixture two or
more things mixed
together that keep
their own properties
(page 340)

solution a kind of
mixture with parts
that do not easily
separate (page 343)

Matter Changes

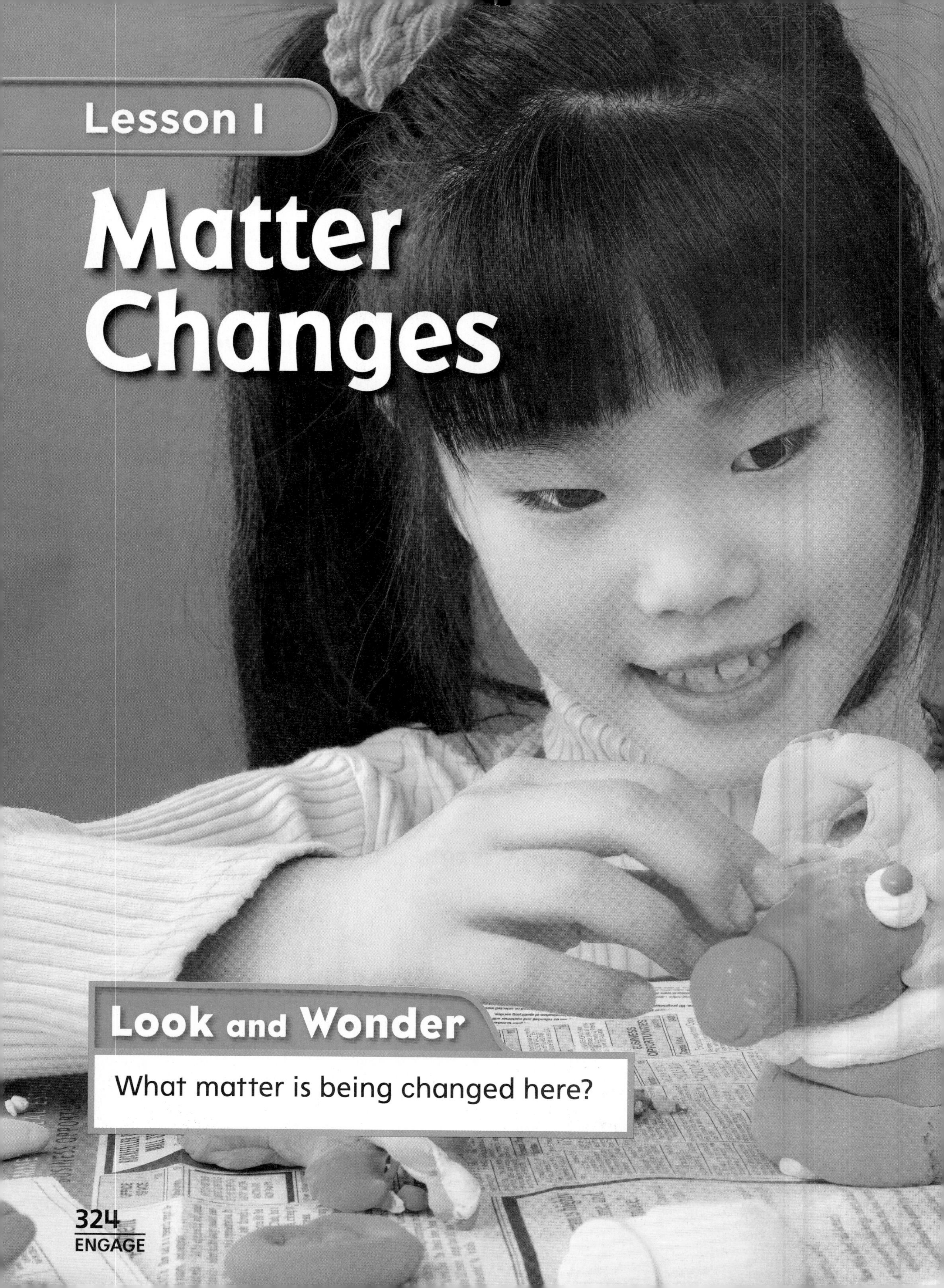

Look and Wonder

What matter is being changed here?

How can clay be changed?

What to Do

1. **Measure.** Find two pieces of clay that are the same mass. Use a balance to show they are equal.

2. **Squeeze** and shape one piece of clay into a ball. Describe its properties.

3. **Predict.** Do you think the mass of the clay changed after it was made into a ball? Place it back on the balance to find out.

4.  **Be Careful!** Cut the clay ball into two halves with a plastic knife. Make the two pieces into two figures.

5. **Draw Conclusions.** How did you change the clay?

Explore More

6. **Investigate.** What other ways can you change clay? Will the mass change?

You need

modeling clay

balance

plastic knife

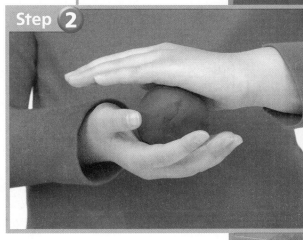

Step 2

What are physical changes?

Matter can change in different ways. You can change the size or shape of matter. This is called a **physical change**.

When you cut, bend, fold, or tear matter, you cause a physical change. You can change the shape or size of paper by cutting or folding it. It is still paper. Its properties are the same.

◀ **Folding and writing on paper are physical changes.**

When you only change the shape of matter, its mass stays the same. ▼

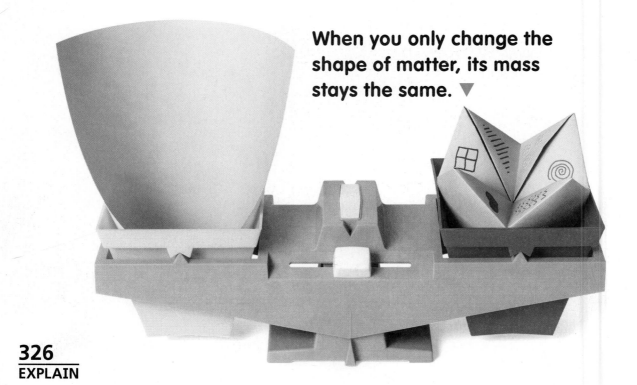

Sometimes, the temperature of matter changes. On a cold day, water can change to ice. This is a physical change.

Wetting and drying can be physical changes, too. Wet mud looks and feels different from dry mud.

▲ **The water on this branch has changed to ice.**

 What is a physical change you could make to juice?

The color of dry mud is different from wet mud. Mud feels squishy when it is wet, and hard when it is dry.

What are chemical changes?

Sometimes the properties of matter can change. This is called a **chemical change**. When matter goes through a chemical change, sometimes it is not easy to change it back. It becomes a new kind of matter with different properties.

When you burn paper, you can not change it back. Seeing light and feeling heat are clues that a chemical change may be happening. All matter does not change in the same way.

≡Quick Lab

Observe a slice of apple. Infer what causes the apple to go through a chemical change.

Chemical Changes		
Before	**After**	**Cause**
		Heat causes the matchstick to burn. The properties of the matchstick have changed.
		Water and air can cause metal to rust. Rust is a chemical change that happens slowly.
		Water and air do not change the properties of plastic.

Read a Chart

How did the metal nail change?

Heat causes the egg to go through a chemical change that you can see and smell.

✓ How can you tell if a chemical change has happened?

Think, Talk, and Write

1. **Problem and Solution.** Describe how you could keep a bicycle from rusting.

2. What are three examples of physical changes?

3. What happens to a banana peel over time? Write about it.

Math Link

Does the mass of an object change when you fold the object? How could you find out?

LOG ON e-**Review** Summaries and quizzes online at **www.macmillanmh.com**

Changes of State

Kilauea Volcano, Hawaii

Look and Wonder

Volcanoes are so hot that rocks can melt and flow like a liquid. How else can heat change things?

How can heat change matter?

What to Do

① **Predict.** What do you think will happen to butter and chocolate in sunlight?

② **Observe.** Place the butter and chocolate on two plates. Draw how they look.

③ **Predict.** How will the Sun's heat change each thing? Find a sunny spot. Leave the plates in the sunlight.

④ **Communicate.** What happens to each thing after one hour? Draw how they look. Compare your pictures.

Explore More

⑤ Now try another item. How will it change?

You need

paper plates

butter

chocolate

Step **②**

How can heating change matter?

Have you ever left a bar of chocolate in your pocket in summer? When you reached in to get it, chances are it was melting.

Melting means changing from a solid to a liquid. Some solids, like gold and glass, will only melt when they are very hot. Other solids, like ice and butter, melt at much lower temperatures.

◄ When gold melts, you can pour it into molds. As the gold cools, it will harden.

▼ Solid ice cubes melt when left at room temperature.

Water can change to a gas when it is heated. **Evaporate** means to change from liquid to gas and go into the air.

If enough heat is added to water, it will boil. When water boils, you can see bubbles. The bubbles show that the water is changing to a gas called water vapor. We can not see water vapor.

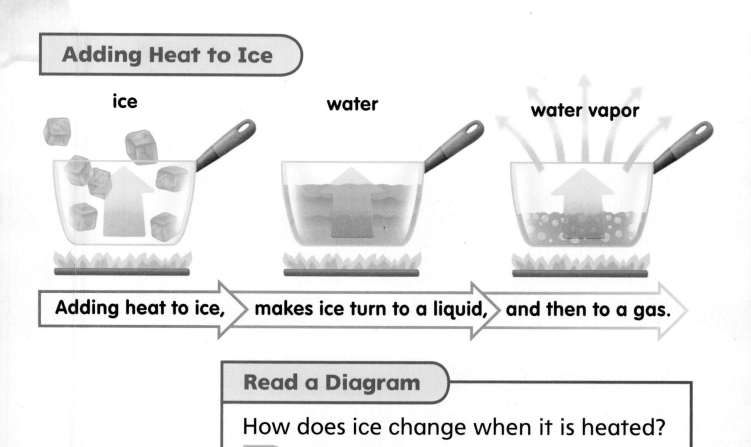

Adding Heat to Ice

ice   water   water vapor

Adding heat to ice, ⟩ makes ice turn to a liquid, ⟩ and then to a gas.

Read a Diagram

How does ice change when it is heated?

LOG ON *Science in Motion* Watch what happens when heat melts ice at **www.macmillanmh.com**

 How can heat change solids?

How can cooling change matter?

Matter can also change by cooling, or taking away heat. Gases condense when they are cooled. **Condense** means to change from a gas to a liquid.

Water vapor in the air condenses when it touches cool objects. This is why you see small drops of water on the outside of a cold glass.

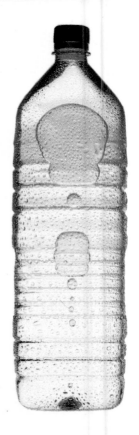

▲ **Water vapor condenses on the outside of a bottle.**

FACT Water that has condensed on a window comes from air inside the room.

When liquids cool, they can freeze, or become solid. Wax and some other liquids will freeze at room temperature. Other liquids, like water, need to be much colder to freeze.

 How does water change when it is cooled?

Quick Lab

Classify pictures of water from magazines as solid, liquid, or gas.

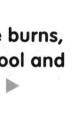

After a candle burns, the wax will cool and become solid. ▶

Think, Talk, and Write

1. **Predict.** What will happen to a puddle of water on a sunny day?

2. What happens when water vapor condenses?

3. Draw and write about how you could change water from a solid to a liquid and back to a solid again.

Math Link

Do you think the mass of ice changes when it melts? How could you find out?

LOG ON **e-Review** Summaries and quizzes online at **www.macmillanmh.com**

Colorful Creations

There are all kinds of colors inside your crayon box. How were those crayons made?

Most crayons are made of wax. This man adds special dye to a tub of wax to give the wax color.

The colored wax is melted into a liquid. Then a worker pours this hot wax into a mold.

Inside the mold there are hundreds of holes shaped like crayons. The wax fills each hole. Then the mold is cooled with cold water.

This machine packs the crayons into boxes.

This woman checks the crayons by hand to make sure they are good.

Talk About It

Predict. What will happen if the mixture of wax is left out at room temperature?

▲ Now the crayons are ready to be shipped to a store near you!

AMERICAN MUSEUM OF NATURAL HISTORY

Mixtures

San Francisco, California

Look and Wonder

What does a mixture of sand and water feel like? How is it different from dry sand?

What mixes with water?

What to Do

1. **Measure.** Add $\frac{1}{4}$ cup salt to one cup of water. What happens?

2. **Measure.** Add $\frac{1}{4}$ cup sand to another cup of water. Does the sand change?

3. **Compare.** Stir both mixtures with a spoon. Let them sit. What happens? How are the mixtures different from each other?

Explore More

4. **Investigate.** Tell how you could take the sand and the water apart. Can the salt be taken out of the water?

You need

measuring cup

2 plastic cups

2 spoons

salt

Step 2

salt   sand

What are mixtures?

When you put salt into water, you make a mixture. A **mixture** is two or more things put together. Mixtures can be any combination of solids, liquids, and gases.

When you glue different things to paper, you make a mixture. When you put pieces of clay together, you also make a mixture.

Papier mâché is a mixture of flour, water, and newspaper.

You can cover items with wet newspaper to make things.

Sometimes when you mix things together, it is easy to pick them apart again. You can see the different parts of the mixture. The things in the mixture do not change.

This pencil holder used papier mâché and buttons. What else do you see in the mixture?

 What kinds of matter can be used to make a mixture?

Which mixtures stay mixed?

Sometimes when you mix things, it is not easy to change them back. When you make a shake or smoothie, you mix different foods together. It is hard to take apart after it has been blended.

before

after

Read a Photo

Which mixture is harder to take apart?

A **solution** is a mixture that is hard to take apart. Sugar and water make a solution. Sugar will **dissolve**, or stay evenly mixed in the water.

Sand and water can be mixed, but they do not make a solution. The sand does not stay mixed and sinks to the bottom of the glass.

The drink mix dissolves in the water. ▶

▲ The soapy water is a solution. The dishes are a mixture.

 How is a solution a special kind of mixture?

How can you take mixtures apart?

Have you ever picked the pretzels out of a snack mix? You were taking apart a mixture. Some mixtures are harder to take apart.

Filters are screens that trap solids but let liquids flow through. Magnets can also help take some mixtures apart. They can be used to separate mixtures with iron or some other metals.

Quick Lab

Investigate how evaporation helps take a mixture apart.

A magnet can help take a mixture of sand and iron filings apart.

A filter can help take a mixture of sand and water apart.

Some mixtures can be even harder to take apart. Evaporation can be used to take a solution of salt water apart. If you leave salt water out to dry, the water evaporates. The salt is left behind.

Water has evaporated from the ocean and left salt here.

✓ How do filters help separate mixtures?

Think, Talk, and Write

1. **Main Idea and Details.** Describe how different things mix with water.

2. How can you take apart a solution of salt and water?

3. Write about how you could take a mixture of paper clips and toothpicks apart.

Health Link

What foods are mixtures? Look for food mixtures in books, magazines, or at the grocery store. Make a list.

Writing a Recipe

A recipe is a set of directions for making something. The steps are explained in order. A recipe can tell you how to make mixtures by adding things together.

Trail Mix Recipe

½ cup peanuts

½ cup raisins

½ cup chocolate chips

½ cup sunflower seeds

Write About It

You can write a recipe! Explain how you would use some of the fruit here to make a fruit salad. Tell why it is a mixture.

Remember

When you write to explain, you tell how to do something. You write the steps in order.

LOG ON **e-Journal** Write about it online at **www.macmillanmh.com**

Muffin Math

Maria and her dad are making muffins. They want to know how much flour they have to buy.

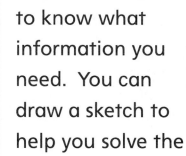

Solve a Problem

The recipe says they will need 2 cups of flour to make 12 muffins.

Maria and her dad want to make 24 muffins. How much flour will they need?

Remember

Read carefully to know what information you need. You can draw a sketch to help you solve the problem.

How Things Change

All around me, matter changes. Cutting is a physical change. Even though there are more pieces, this is still an apple. Physical changes do not change what something is.

All around me, matter changes. Batter cooks in the oven and becomes a muffin. This is a chemical change. The batter has new properties now.

All around me, matter changes. I can make a mixture of chocolate powder and milk. The chocolate milk stays mixed. It is a solution.

All around me, matter changes. I can freeze juice. When something freezes, it changes to a solid.

351

Vocabulary

Use each term once for items 1–6.

1. When wood burns, there is a _____.

2. Water in the air can _____ or change into a liquid.

3. Sugar and water form a mixture that will stay mixed. It is called a _____.

4. Fruit salad is a kind of _____.

5. Tearing paper is a _____.

6. After the snowman melts, the water will turn to a gas, or _____.

chemical change

condense

evaporate

mixture

physical change

solution

Answer the questions below.

7. **Communicate.** Which photo shows a physical change? Which shows a chemical change? What are some other examples of each kind of change?

8. **Predict.** What will happen if ice is heated at a high temperature for a long time?

9. Describe how a solution of sugar and water is different from a mixture of sand and water.

10. What are some ways to separate mixtures?

11. How can matter change?

Food Chemist

Would you like to make your own cereal or flavor of juice? You could become a food chemist. Food chemists explore ways to make new and more delicious foods.

Food chemists learn how to make yogurt smooth. They find out how to keep cereal crunchy. They might find a way to freeze vegetables so they taste fresher. Food chemists have to understand the science of how food products are made.

food chemist

More Careers to Think About

nutritionist

chef

LOG ON e-Careers at www.macmillanmh.com

Motion and Energy

Roller coasters can go more
than 100 miles per hour.

Echolocation

Our sense of hearing is very helpful, but does it help us eat? For bats, it does! Bats are able to use their ears and echolocation to find food.

Since bats live in dark caves and cannot see very well at night, they need to use sounds to move around and catch their next meal.

Bats make a high-pitched squeak when they use echolocation. The sound makes a vibration, which can bounce off of a moth, and then bounce back to the bat. Once the bat can sense where the moth is, it can fly in and grab its dinner!

Talk About It

How does sound vibration help animals get food?

How Things Move

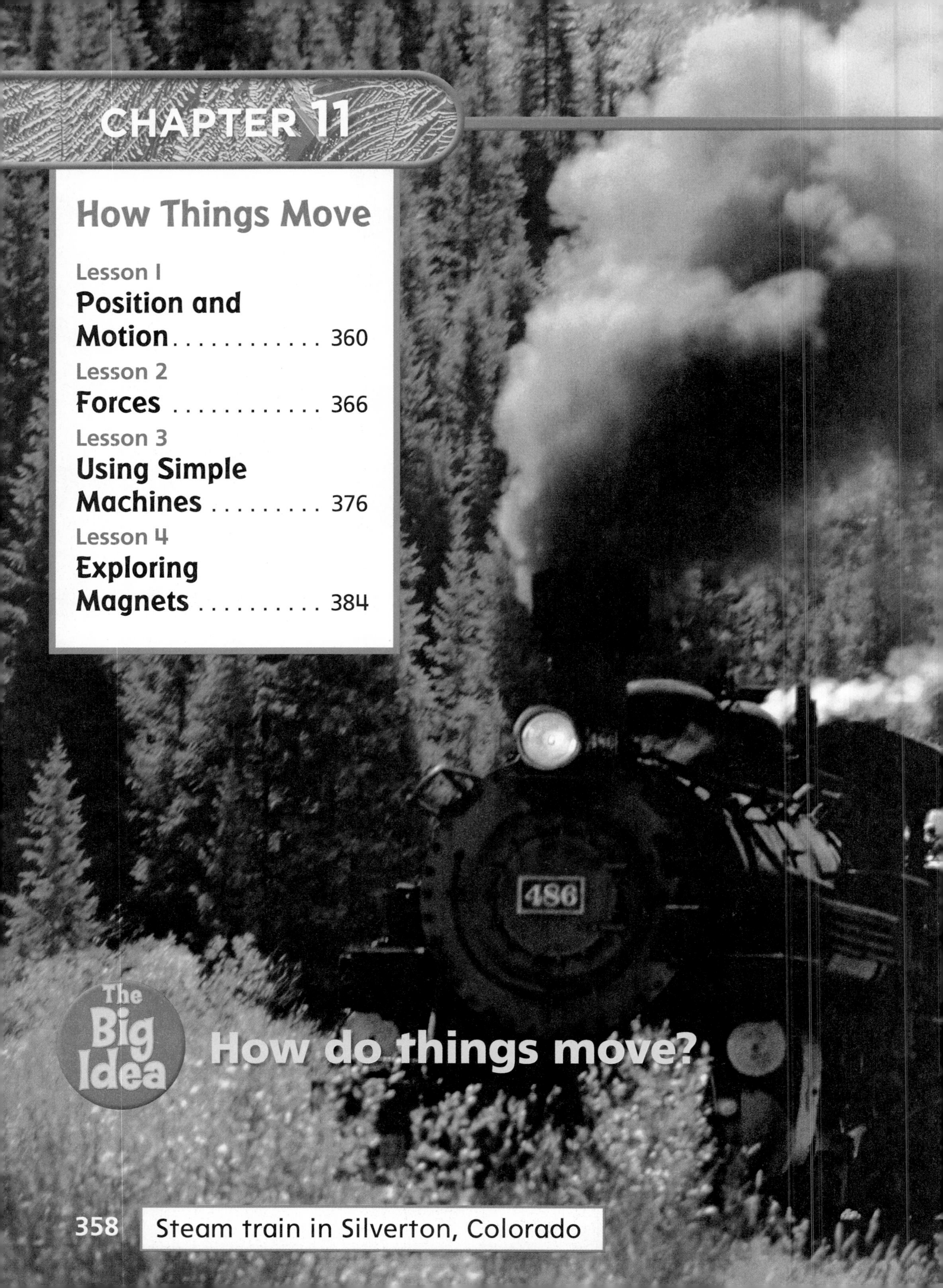

The Big Idea How do things move?

Key Vocabulary

motion a change in the position of an object (page 363)

friction a force that slows down moving things (page 371)

lever a simple machine made of a bar that turns around a point (page 378)

repel to push away or apart (page 388)

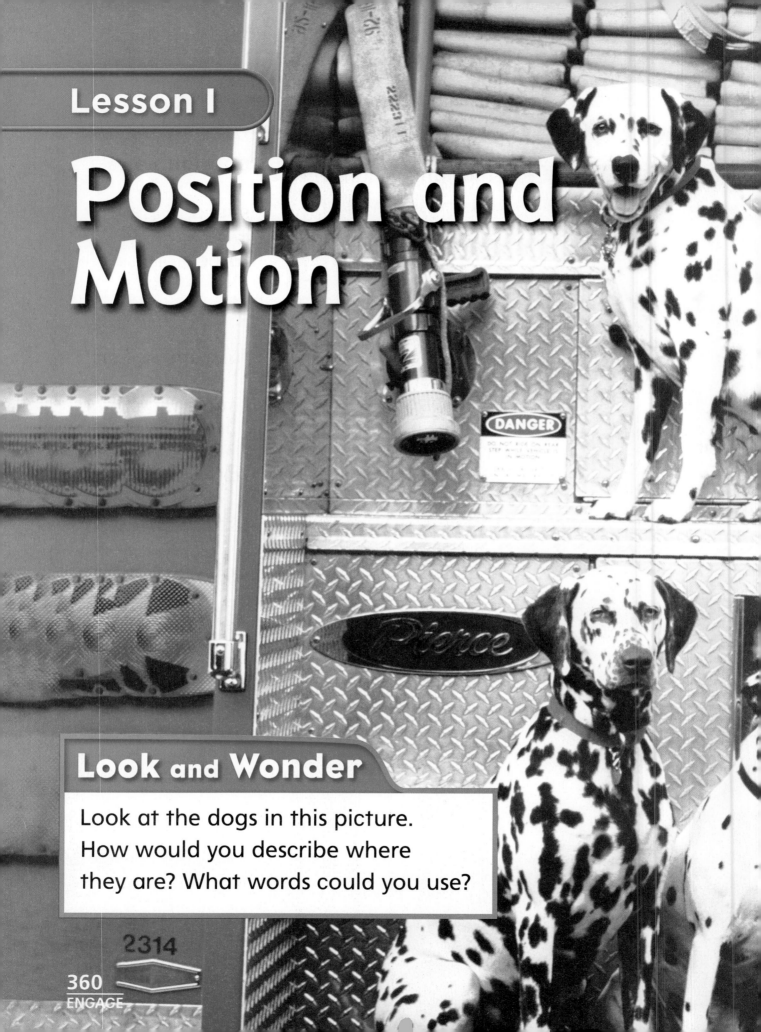

Position and Motion

Look and Wonder

Look at the dogs in this picture.
How would you describe where
they are? What words could you use?

What words help us find things?

What to Do

1. Work with a partner. Pick an object in the classroom. Do not tell your partner what the object is.

2. **Communicate.** Describe where your object is. Give clues to your partner. Ask your partner to find the object.

3. Switch with your partner and try again.

4. **Draw Conclusions.** Which words in your description were most helpful to your partner?

Explore More

5. **Communicate.** Draw a picture and write directions to find an object in your picture. Then switch with a partner.

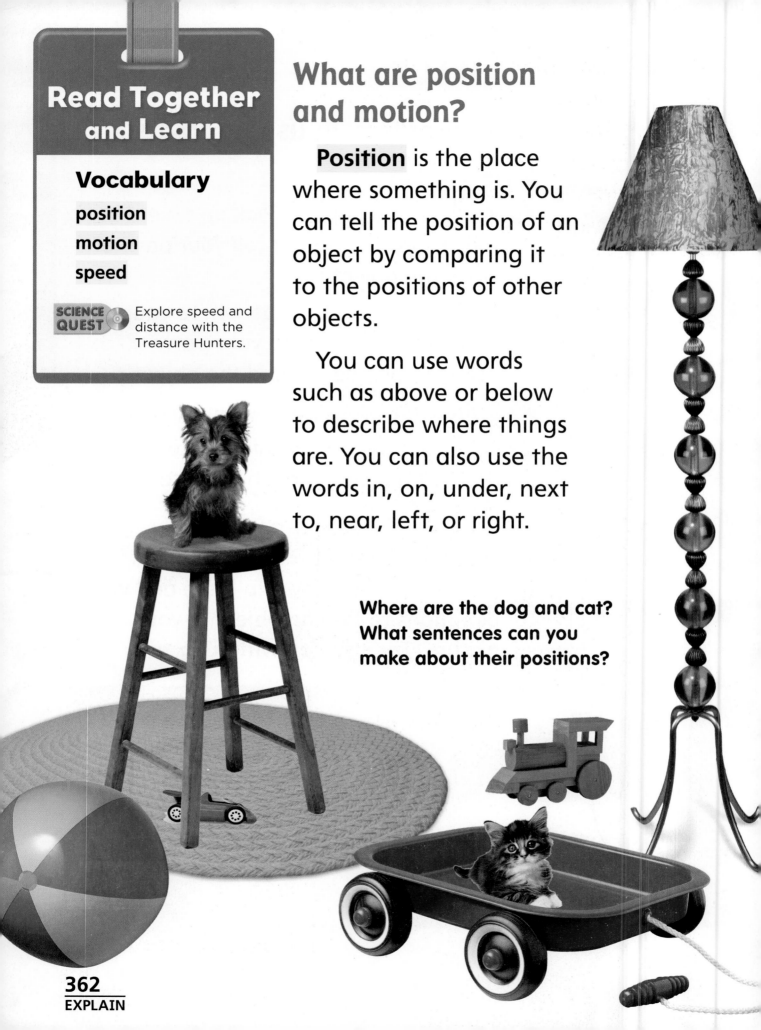

Vocabulary

position

motion

speed

SCIENCE QUEST Explore speed and distance with the Treasure Hunters.

What are position and motion?

Position is the place where something is. You can tell the position of an object by comparing it to the positions of other objects.

You can use words such as above or below to describe where things are. You can also use the words in, on, under, next to, near, left, or right.

Where are the dog and cat? What sentences can you make about their positions?

When something moves, its position changes. **Motion** is a change in the position of an object. Some ways objects move are up, down, around, sideways, or zigzag. You can describe an object's motion by telling how its position changed.

✓ How can you describe where an object is and how it moves?

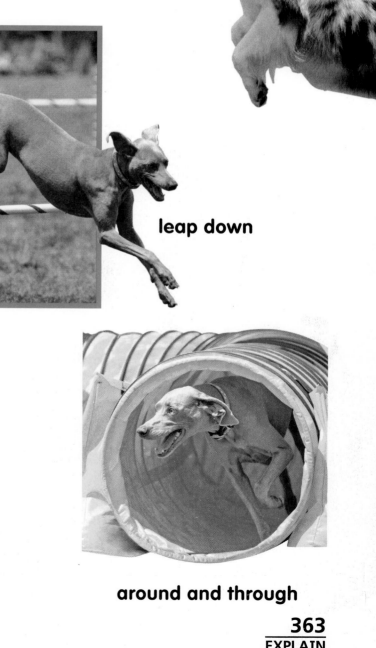

jump up

leap down

zigzag

around and through

What is speed?

Some things, such as snails, move slowly. Others, such as cheetahs, move quickly. **Speed** is how far something moves in a certain amount of time.

Quick Lab

Measure three meters in the classroom. Walk and then hop the distance. Record the time it took for each.

A cheetah's speed can be measured with a stopwatch and a tape measure. The fastest objects move farthest in a certain amount of time.

Animal Speed

zebra

cheetah

lion

animal

miles per hour

0   10   20   30   40   50   60   70

Read a Graph

Which animal is the fastest?

 What are some objects that move at high speeds?

Think, Talk, and Write

1. **Sequence.** You walk from your desk to your teacher's desk. Describe the order of objects you would pass as you move.

2. What are some words that you can use to describe motion?

3. Write about what speed is.

Social Studies Link

Make a map of your classroom. Draw yourself on the map and describe your position.

LOG ON **e-Review** Summaries and quizzes online at **www.macmillanmh.com**

Forces

Look and Wonder

How can you make something move?
How can you make it move farther?

How do you make things go farther and faster?

What to Do

1. Line up the car at a starting line. Push the car gently over the line.

2. **Measure.** How far did it go?

3. Do the activity again, but this time push the car harder. Observe what happens.

Explore More

4. **Predict.** What might happen if you pulled the car toward you with your hands? Would it go as far?

You need

toy car

masking tape

ruler

Step 2

Vocabulary

force

gravity

friction

SCIENCE QUEST ○ Explore pushes and pulls with the Treasure Hunters.

What makes things move?

Objects can not start to move on their own. You have to use a push or a pull to put something in motion.

When you play soccer, you kick the ball to move it across the field. Your kick is a push. If you do not kick the ball, it will stay in the same place.

A stronger kick will make an object move farther.

A push or pull is called a **force**. If you push something, it will move away from you. If you pull it, it will move closer to you.

A kick is a kind of push. Opening a drawer is a kind of pull. You can move different objects with different amounts of force.

 Why do we need forces?

What is making the cart move?

Both groups are pulling the rope. Why does it not move?

What are some forces?

When you let go of a ball, it falls. **Gravity** is a force that pulls down on everything on Earth. When you jump in the air, gravity pulls you back down to the ground. Gravity pulls on objects through solids, liquids, and gases. The amount of force that pulls something down toward Earth is called its weight.

Why is the ball falling? What do you think will happen to the dog?

FACT All planets have gravity.

When you skate, you drag a rubber stopper on the ground to stop. The dragging causes friction. **Friction** is a force that slows down moving things. Friction happens when two things rub together.

There is usually more friction on rough surfaces than on smooth ones. It is usually harder to push or pull something on a rough surface than on a smooth surface.

✔ How are gravity and friction alike?

Quick Lab

Slide a wooden block on different slanted surfaces. Compare how friction affects the speed of the block.

Dragging the rubber stopper on the ground causes friction. This slows the skater down.

The ball falls to the grass and rolls. Friction makes the ball slow down and stop. ▼

How can forces change motion?

You know that forces can change how things move. Forces can make things start moving, speed up, slow down, and stop. Forces can make things change direction, too. In softball, the players use forces to change the direction of the ball.

 Think about a sport that uses a ball. How does the ball change direction?

The pitcher uses a force to throw the ball toward the batter. ▼

How a Ball Changes Direction

◄ The batter hits the ball with a push. It changes direction and flies toward the outfield.

The player in the outfield catches the ball and uses a force to stop its motion. He can also use a force to throw the ball to another player. ▶

Read a Diagram

What kind of forces do the players use?

LOG ON *Science in Motion* Watch forces work at **www.macmillanmh.com**

Think, Talk, and Write

1. **Cause and Effect.** What happens when you put more force on an object?

2. Why is it hard to push something on some surfaces?

3. Write a story about a day without gravity.

Social Studies Link

Learn about a sport played in another country. Describe the pushes and pulls in this sport.

LOG ON e-Review Summaries and quizzes online at **www.macmillanmh.com**

Meet Héctor Arce

Héctor Arce is a scientist at the American Museum of Natural History. Héctor studies how stars form. When gravity pulls together huge clouds of gas and dust, stars form. Gravity makes their centers so hot that they light up. This is why stars shine in our night sky.

Gravity is the force that keeps you on Earth. You may not be able to see gravity, but it is all around you. In fact, it is everywhere! There is gravity on planets, moons, and stars. How powerful is gravity? It is powerful enough to create a star!

Héctor Arce is an astrophysicist, a scientist who studies the planets, moons, and stars.

Héctor uses a telescope like the one in this building to get a closer look at stars.

Talk About It

Cause and Effect.
How do stars form?

AMERICAN MUSEUM OF NATURAL HISTORY

375
EXTEND

Using Simple Machines

Look and Wonder

Have you ever used a shovel?
How does it make digging easier?

Which side will go up?

What to Do

1. Tape a marker to the middle of your desk.

2. Tape 10 pennies to the edge of one end of a ruler. Tape 5 pennies to the edge of the other end.

3. **Predict.** What will happen if you put the middle of the ruler on the marker? Which side will lift up? Try it. Was your prediction correct?

Explore More

4. Try to move the ruler so that 5 pennies can lift 10 pennies. Where did you need to move the ruler?

You need

marker

tape

ruler

15 pennies

Step **2**

Read Together and Learn

Vocabulary

simple machine

lever

fulcrum

ramp

What are levers and ramps?

A **simple machine** is a tool that changes the size or direction of a force. A simple machine can make work easier.

A **lever** is a bar that moves against an unmoving point. The unmoving point that a lever moves against is called a **fulcrum**. Shovels and seesaws are levers. When you push down on one side of the lever, the other side moves up.

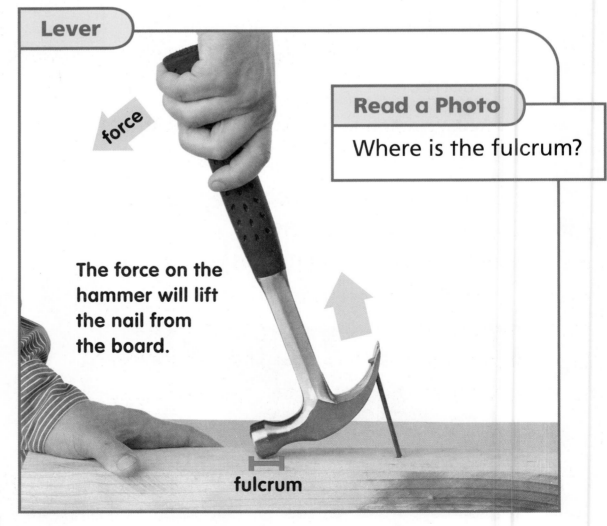

Lever

force

The force on the hammer will lift the nail from the board.

fulcrum

Read a Photo

Where is the fulcrum?

**Pushing a heavy box up a ramp
needs less force than lifting the box.**

Another kind of simple machine is a ramp. A **ramp** is a surface that is straight and slanted. Ramps can be used to move an object from one place to another. Pushing something up a ramp is easier than lifting it. Less force is needed to move something on a long, low ramp than on a short, steep ramp.

 How do a lever and ramp make work easier?

What are other simple machines?

A bicycle uses a simple machine called a wheel and axle. A wheel and axle is made of a wheel and a bar, or axle. The bar is connected to the center of the wheel. When the wheel turns, the bar turns too.

A doorknob and a steering wheel use a wheel and axle. Each axle on a car or bus has two wheels attached.

Where is the axle on this monster truck?

≡Quick Lab

Investigate how to make a pulley. Use the pulley to lift a pail filled with blocks.

A pulley is also a simple machine. A pulley is made with a rope that moves around a wheel. When you attach a pulley to a object, you can change the direction of the force on the object. A pulley can help lift an object up high.

 When might it be helpful to use a pulley?

When you pull the rope down, the pig in the pail goes up to the pulley.

Think, Talk, and Write

1. **Summarize.** How can simple machines help you?

2. What are some kinds of simple machines?

3. Write about a simple machine used in your home.

Math Link

Make a tally chart of simple machines used at home and school.

LOG ON **e-Review** Summaries and quizzes online at **www.macmillanmh.com**

Slip and Slide

Have you ever walked on ice? It is smooth and slippery! Sometimes penguins slide on their bellies to move.

Write About It

Explain why penguins can slide on the ice. Think about what you learned about forces. Make sure to explain why ice is slippery.

Remember

When you write to give information, you give facts.

LOG ON **e–Journal** Write about it online at **www.macmillanmh.com**

How Far Did It Move?

These students are playing softball.
They want to know how far the ball moved.

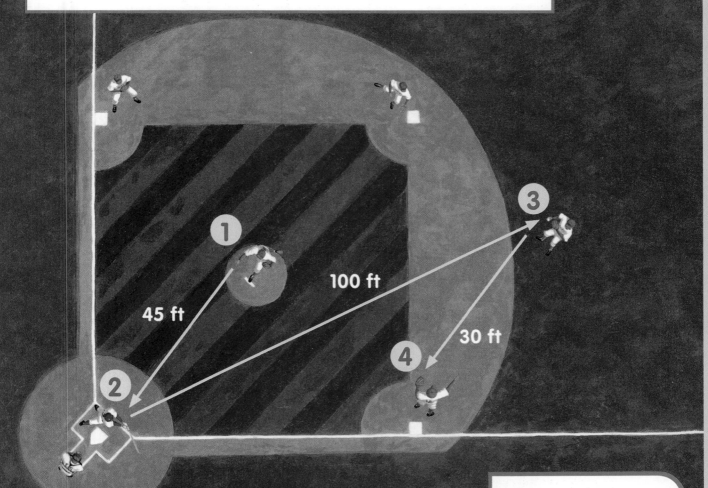

1

100 ft

45 ft

3

30 ft

4

2

Add Measurements

Add the distances the ball moved. How far did it go? How many times did the ball change directions? Make up your own math problem about the softball game.

Remember
Add the 1s first. Then add the 10s. Then add the 100s.

Exploring Magnets

Look and Wonder

Why does the magnet pull some of these objects and not others?

What can a magnet pick up?

What to Do

You need

small objects

1 **Predict.** Put the objects in a bag. Which objects will stick to a magnet?

2 Tie a string to a pencil. Tie a magnet to the end of the string.

paper bag

3 Use the magnet to pull out objects from the bag.

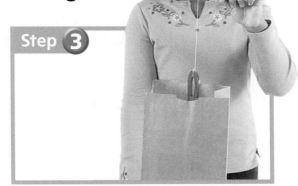

Step 3

string

pencil

Explore More

magnet

4 **Classify.** How are the things that stick to the magnet alike?

What do magnets do?

A magnet can **attract**, or pull, some objects. Magnets attract objects through solids, liquids, and gases. A very strong magnet can pull objects from far away. The farther a magnet is from the object, the weaker the magnet's pull will be.

Many magnets contain iron. Magnets attract objects containing iron. They can also attract objects containing nickel.

The magnet pulls the paper clip without touching it.

Magnets are holding these objects in place.

buy:
— eggs
— milk

There are many objects that magnets can not attract. These include plastic, wood, and some metals. Walk around your classroom with a magnet. See what the magnet attracts and what it does not.

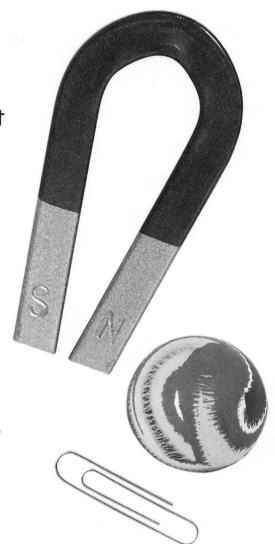

What a Magnet Attracts

object	attracts	does not attract
crayon		✓
screw	✓	
eraser		✓
lock	✓	

Read a Chart

Which objects will stick to a magnet?

✓ Will a magnet attract a button? Why or why not?

What are poles?

The two ends of a magnet are its **poles**. Every magnet has a north pole and a south pole. Put the north pole of one magnet next to the south pole of another. They will attract each other.

Now put the two south poles together. They will **repel** each other, or push apart. The same thing will happen with the two north poles. The push and pull of a magnet is strongest at its poles.

Quick Lab

Cover the labels on two bar magnets. **Investigate** to find which poles are alike and which are different.

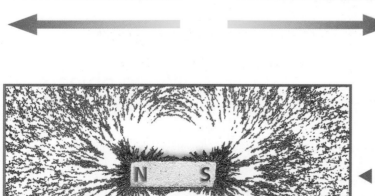

◀ **This magnet attracts tiny pieces of iron.**

FACT Some magnets are much stronger than others.

Earth acts like a big magnet. Like every magnet, it has north and south poles.

A compass is a magnet that is free to spin. The north pole in the magnet points toward Earth's North Pole.

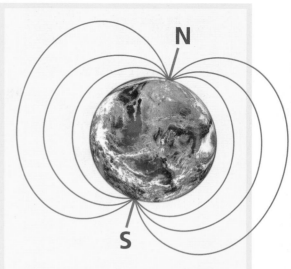

Earth has a magnetic force around the North Pole and South Pole.

The needle inside a compass is a magnet that points to Earth's North Pole.

 Where is the pull of a magnet strongest?

Think, Talk, and Write

1. **Problem and Solution.** Two magnets repel each other. How can you make them stick together?

2. What will a magnet attract?

3. Write about how you can tell if an object might be made of steel or iron.

Art Link

Make a poster that shows how people use magnets.

LOG ON e-Review Summaries and quizzes online at www.macmillanmh.com

Forces Every Day

Push! I use forces all day long. I push the heavy cart to make it move. A push is a force. I pull on the cart to make it stop. A pull is a force, too.

390

Plop! I crack an egg. A force called gravity makes the egg fall into the bowl. Objects move with different motions. I stir the pancake batter around and around.

Flip! I flip my pancakes. Simple machines can make work easier. A lever can change the direction of a force. The force makes the pancake fly through the air.

Yum! The pancake lands on my plate. Other things use pushes and pulls, too. A magnet keeps my shopping list on the refrigerator. It is time to buy more eggs and milk!

393

Vocabulary

Use each word once for items 1–6.

1. When two objects rub together, they can be slowed down by _____.

2. A simple machine that makes it easier to push an object to a higher level is a _____.

3. We can tell where an object is by its _____.

4. Objects fall to the floor because of a force called _____.

5. How far an object moves in a period of time is called _____.

6. A simple machine that moves against a fulcrum is called a _____.

| |
| friction |
| gravity |
| lever |
| position |
| ramp |
| speed |

Answer the questions below.

7. Summarize. Describe the position of the blue paper.

8. Investigate. What can help you move a heavy object?

9. What does gravity do?

10. Describe some of the simple machines in this picture and how they work.

The Big Idea

11. How do things move?

Using Energy

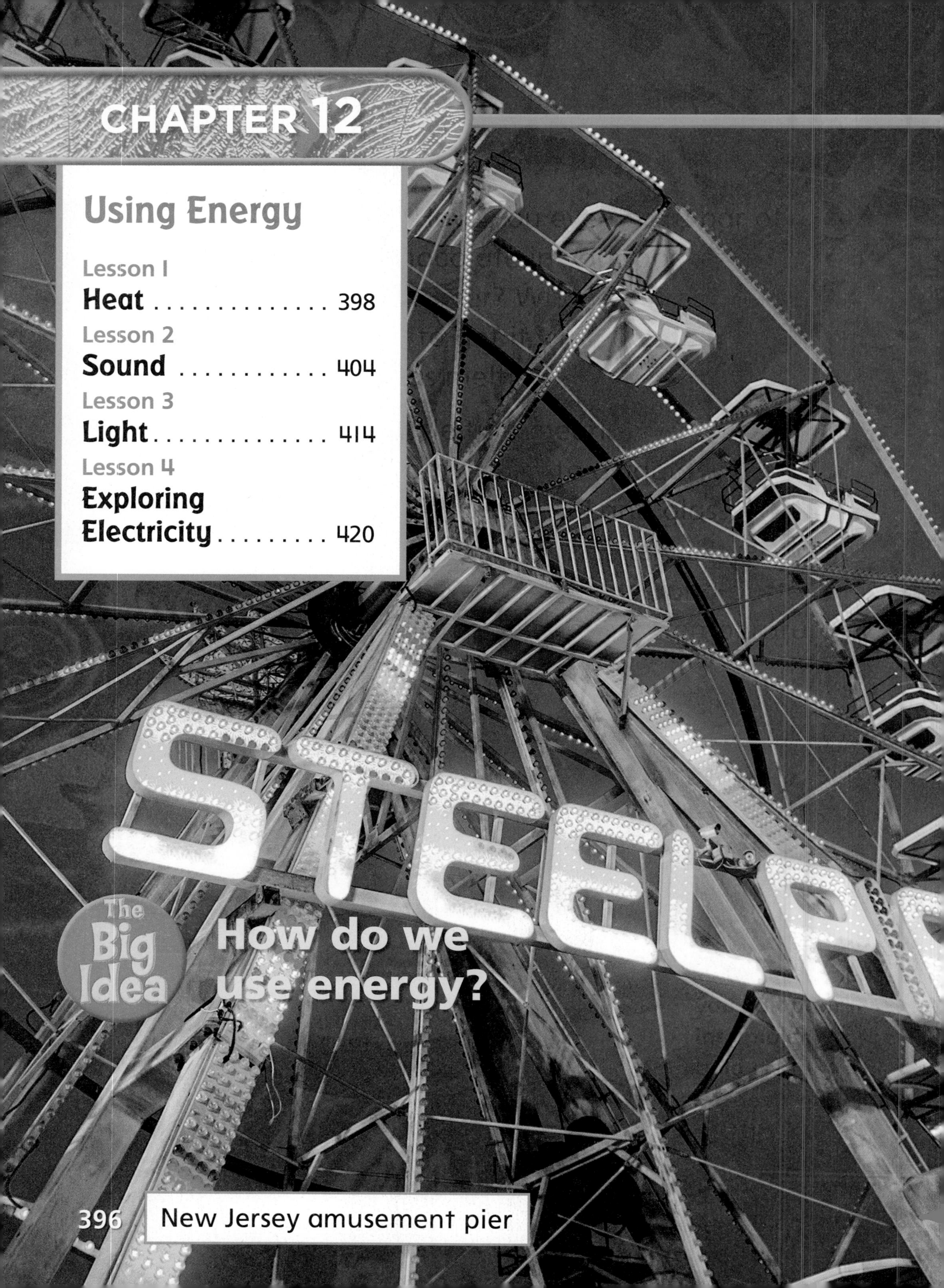

The **Big Idea** **How do we use energy?**

New Jersey amusement pier

Key Vocabulary

More Vocabulary

heat, page 400

sound, page 406

pitch, page 409

light, page 416

reflect, page 416

static electricity, page 424

fuel something that gives off heat when it burns (page 401)

vibrate to move back and forth quickly (page 407)

circuit a path that electricity flows in (page 422)

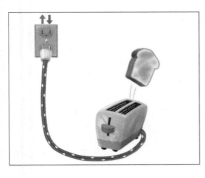

current electricity a kind of energy that moves in a path (page 422)

Heat

Look and Wonder

This is a desert on a sunny day.
How can you tell it is hot?

Where will ice cubes melt more quickly?

You need

ice cubes

2 cups

watch or clock

What to Do

1. Fill two cups with equal amounts of ice. Place one cup in a sunny place. Place the other cup in a shady place.

2. **Predict.** Which cup of ice will melt first?

3. Record how long it takes for the ice in each cup to melt. Why did one cup of ice melt more quickly?

Explore More

4. **Predict.** Put equal amounts of water of the same temperature in two cups. How will each cup of water feel after one hour?

Step 1

sun   shade

Vocabulary
heat
fuel

What is heat?

Energy makes matter move or change. There are many kinds of energy. **Heat** is a kind of energy that can change the state of matter. Heat can turn a solid to a liquid. Heat can turn a liquid to a gas.

We use heat every day. Most heat on Earth comes from the Sun. The Sun warms the air, land, and water on Earth.

On a hot day, the Sun warms the water and land first. Then the air becomes warm.

Heat comes from other things, too. **Fuel** is something that gives off heat when it burns. Gas, oil, wood, and coal can be burned as fuel.

Heat can also come from motion. Rub your hands together quickly to make them warm. Now touch your hands to your face. Heat moved from your hands to your face.

▲ People use fuel to keep warm.

▲ People use fuel to cook food.

✓ How is heat used in your school and home?

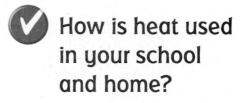

▲ This motion makes heat.

What is temperature?

Temperature is a measure of how hot or cold something is. We measure the temperature of air, water, even our bodies. To measure temperature, we use a tool called a thermometer. Some thermometers have a liquid inside. The liquid goes up or down with the temperature.

Temperature

Read a Photo

Is it hotter during the day or night? How can you tell?

Use a thermometer to **compare** the temperature of soil, water, and air.

✓ What are some things for which you can measure temperature?

Think, Talk, and Write

1. **Main Idea and Details.** Where does most of our heat come from?

2. How do we measure temperature?

 3. Write about some ways people make heat.

Art Link*

Look around your school or home for sources of heat. Draw them.

LOG ON **e-Review** Summaries and quizzes online at **www.macmillanmh.com**

Sound

Look and Wonder

Keep the noise down! How are sounds made? How can some sounds be different from others?

How is sound made?

What to Do

1. Tie the string to the paper clip. Make a hole in the bottom of the cup.

Step 1

2. Pull the string through the hole. The clip keeps the string from pulling through the cup.

3. Wear goggles. Hold the cup and string with a partner. The third partner snaps the string.

4. **Observe.** What happens? How did you make sound?

Explore More

5. **Predict.** How will the sound be different if you change the length of the string? Try it.

You need

string

paper cup

goggles

paper clip

Step 3

Vocabulary

sound

vibrate

pitch

SCIENCE QUEST

Explore sound with the Treasure Hunters.

What makes sound?

Ring! A loud alarm clock wakes you up each morning. How do you hear it? **Sound** is a kind of energy that we can hear.

▲ **When the bells on the alarm clock are hit, they move back and forth quickly.**

How We Hear Sound

▶ **The guitar strings vibrate and make the air around them vibrate.**

Sound energy is made when objects **vibrate**. When an object vibrates, it moves back and forth quickly. When something vibrates, air around the object vibrates also.

The eardrum is the part of our body we use to hear sounds. Messages sent from your ear to your brain tell you what sound you heard.

Quick Lab

Use a tool called a tuning fork. **Observe** what happens when you strike it and place it in water.

How do you hear sounds?

Read a Diagram

How did the sound travel from the guitar to the boy's ear?

 Science in Motion Watch how sound travels at **www.macmillanmh.com**

▼ **These vibrations move to your eardrum so you can hear the sound of the guitar.**

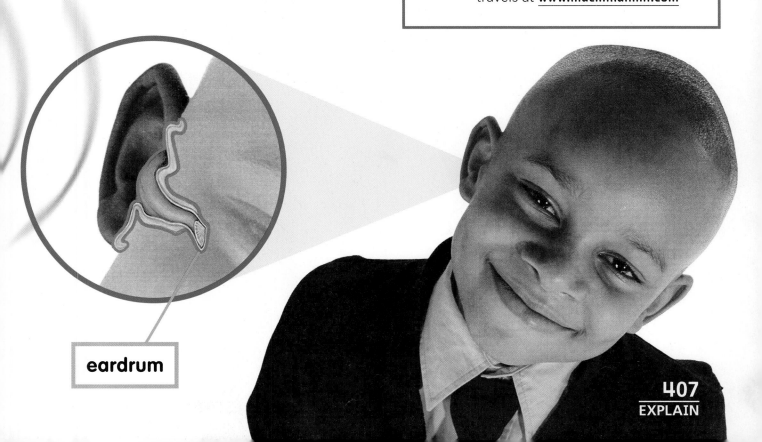

eardrum

How are sounds different?

Not all sounds are the same. You hear loud and soft sounds every day. You can make your voice loud or soft. A whisper has less energy than a shout. Try making loud and soft sounds.

▲ **Small vibrations make soft sounds. The meow of a cat sounds soft.**

▼ **Big vibrations make loud sounds. The roar of a lion sounds loud.**

Pitch is how high or low a sound is. Fast vibrations make sounds with a high pitch. Slow vibrations make sounds with a low pitch.

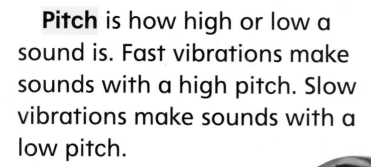

▶ If you snap a short, tight string, it makes a high pitch.

▶ If you snap a long, loose string, it makes a low pitch.

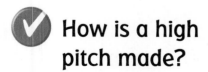 How is a high pitch made?

harp

What do sounds move through?

Place your ear against your desk. Now gently tap the desk with your pencil. You hear vibrations through the wood of the desk. Sound moves through solids, such as wood or plastic.

Sound moves through liquids also. Have you ever heard sound under water? The water vibrates and you hear sound.

▼ **Dolphins and other animals make sounds under water to communicate with each other.**

Most sounds you hear move through air. Air is made of gases. The closer you are to a sound, the louder it sounds. The farther you are from a sound, the softer it sounds.

▲ How can you tell when a fire engine is close by or far away?

What can sounds move through?

Think, Talk, and Write

1. **Problem and Solution.** How would you get a guitar string to make a sound with a high pitch?

2. Why do your hands make a sound when you clap them together?

 3. Write about how you would make a sound louder.

Music Link

Make your own musical instruments. Stretch rubber bands around a plastic cup. Vibrate the rubber bands to make different pitches.

LOG ON **e-Review** Summaries and quizzes online at **www.macmillanmh.com**

Sound Off!

Think about the sounds you hear every day. Some sounds are loud and others are soft. Some sounds are high and others are low.

✏ Write About It

Describe the pitch and volume of a sound you hear every day. How do we use sounds? Why are sounds important?

Remember

When you describe something, you give details.

 e –Journal Write about it online at **www.macmillanmh.com**

Drum Fun

Miss Lee sells four different drums in her store. The first drum is 10 centimeters wide. The second drum is 20 centimeters wide. The third drum is 30 centimeters wide.

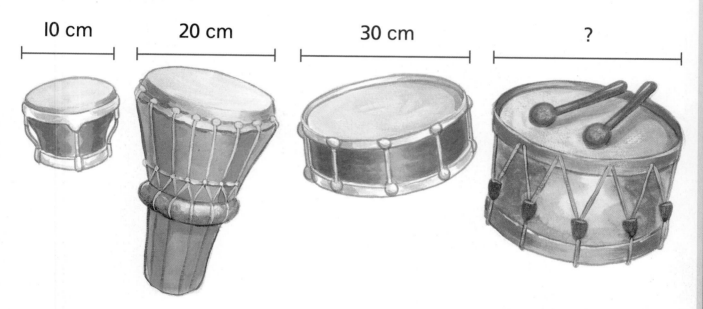

10 cm   20 cm   30 cm   ?

Follow the Pattern

How wide is the fourth drum?
Follow this number pattern:

10 + 10 = 20

20 + 10 = 30

30 + ? = ?

Miss Lee knows that the smallest drum has the highest pitch. Which drum has the lowest pitch?

Remember
You can use a pattern to help you solve problems.

Light

Look and Wonder

Where is this light coming from?
What is blocking some of the light?

What does light pass through?

What to Do

① **Predict.** Which materials will light pass through? Which will block the light?

② Work with a partner. Hold up the cardboard. Hold plastic wrap three inches in front of the board. Your partner shines the flashlight on the object.

③ **Observe.** Did the plastic wrap block the light or did the light pass through it?

④ **Compare.** Which objects block the light and which let light pass through?

Explore More

⑤ **Predict.** What might happen with other classroom items? Try it.

You need

flashlight

cardboard

plastic wrap

various items

Step ②

What is light?

You need light to see things. **Light** is a kind of energy. You see things because light will **reflect**, or bounce off things around you. Light that reflects off objects enters your eyes. Then you can see the objects.

Some sources of light are the Sun, lightbulbs, and flashlights. Most light on Earth comes from the Sun.

Smooth, shiny objects such as mirrors reflect a lot of light.

Have you ever made a shadow on a wall? A shadow is a dark area where light does not reach.

Different objects let different amounts of light through. A book is a solid object. It can block light and make a shadow. Glass is clear. It does not make a shadow because light passes through it.

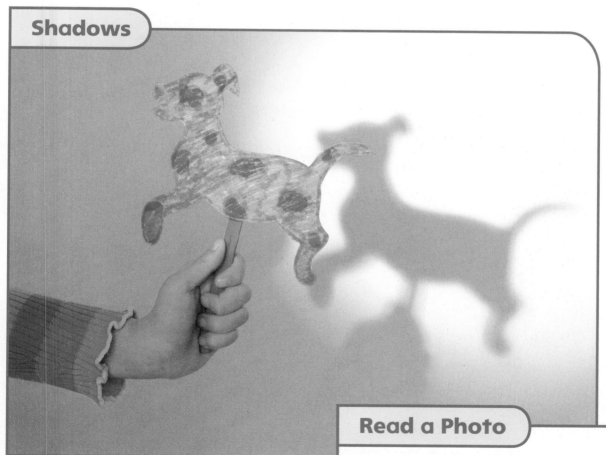

Shadows

Read a Photo

How is this shadow made?

 What are some sources of light?

How do we see color?

Did you know light can bend? Light is a mix of all colors. When white light bends, it separates into different colors. Then we can see the colors of the rainbow.

A prism is an object that can make light bend.

≋Quick Lab

Use a prism and sunlight to see the colors of the rainbow. **Observe** and draw what you see.

prism

Raindrops can act like prisms and separate light to make rainbows.

Have you ever seen colored lights? A filter is a tool that lets only certain colors of light pass through it.

Some filters let only one color pass through. A red filter blocks all colors except red. You see only red light with a red filter.

Colored glass makes a white light look red, green, or yellow.

 What color is most light we see?

Think, Talk, and Write

1. **Sequence.** What happens when we see objects?

2. What kind of objects make shadows?

3. Write a list of things that light can not pass through.

Art Link

Make a filter. Cover a flashlight with colored plastic wrap. Then make shadow puppets!

Exploring Electricity

The Bay Bridge in San Francisco, California

Look and Wonder

How do you think these lights get their energy?

What makes the bulb light up?

What to Do

① **Predict.** Look at the battery, bulb, and wires. How could you put them together to light the bulb? Record your ideas with a partner.

② △ **Be Careful.** Try your ideas. Which of your ideas made the bulb light? Which ideas did not work?

③ **Record Data.** Write down your results with your partner. How many ways did you make the bulb light up?

Explore More

④ **Predict.** How could you make a second bulb light up? What else would you need?

You need

wire

battery

lightbulb

Step **②**

Vocabulary

current electricity

circuit

static electricity

What is current electricity?

Do batteries make some of your toys work? Batteries make a kind of electricity. **Current electricity** is a kind of energy that moves in a path. The electricity moves along a path called a **circuit**. The circuit needs to be completely connected for the electricity to move.

Circuit

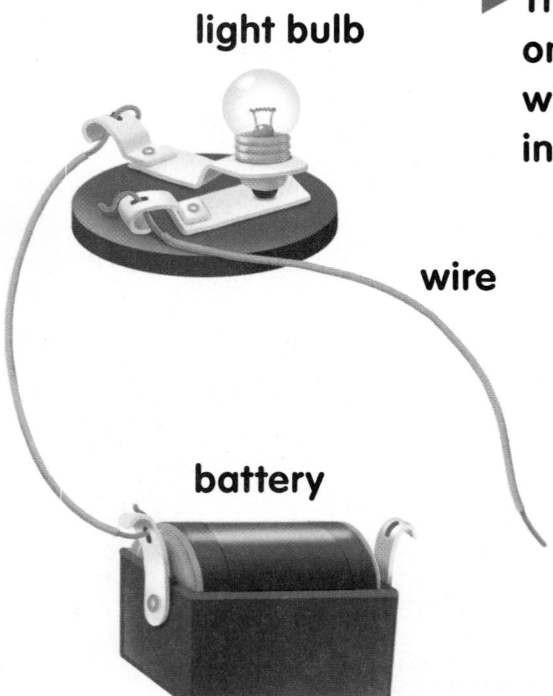

light bulb

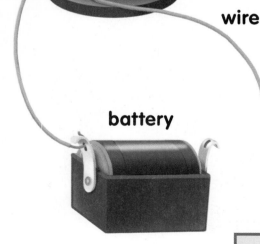

wire

battery

▶ **The light goes on only when all the wires are connected in a complete circuit.**

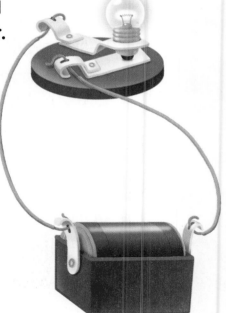

Read a Diagram

Which picture shows a complete circuit?

Current electricity can be changed into heat, light, or sound energy. It can also make things move. Current electricity can come from batteries or from outlets in the wall.

Buildings called power plants change other kinds of energy into electricity. The electricity runs through wires into your house and into the outlets. When you plug in your toaster and turn it on, you complete the circuit with the power plant.

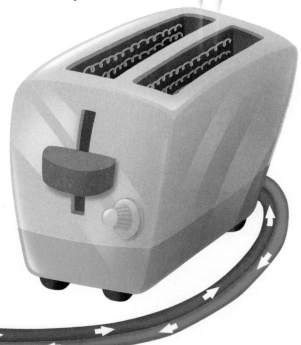

► **Electricity can move through the circuit when the toaster is plugged in.**

 How do you use current electricity every day?

FACT ▷ Electricity comes from power plants through wires, not from your wall.

What is static electricity?

You take your clothes out of the dryer. They are stuck together! This happens because of static electricity.

Static electricity is a kind of energy made by tiny pieces of matter. You can not see these pieces of matter, but they are everywhere.

When tiny pieces of matter attract or repel each other, they have a charge.

Lightning is static electricity. Charges made in a storm jump between the clouds and the ground.

Like a magnet, some of these pieces of matter stick together. Others push apart.

Charges can build up on one object and jump to another. Sometimes you can see or hear a static charge move from one object to another.

The girl's hair is attracted to the charged balloon, so it sticks up.

 What are some examples of static electricity?

Think, Talk, and Write

1. **Cause and Effect.** How does a battery make your toy work?

2. What kind of energy causes your socks to stick together?

3. Write about what your day would be like without electricity.

Social Studies Link

Research and write about how people use electricity.

It's Electric

You can flip a switch to turn on a light, a computer, or a dishwasher. They all use electricity.

The electricity starts at a power plant. At the plant, energy turns a large wheel called a turbine. The energy might come from burning coal or oil, flowing water, wind, or nuclear reactions.

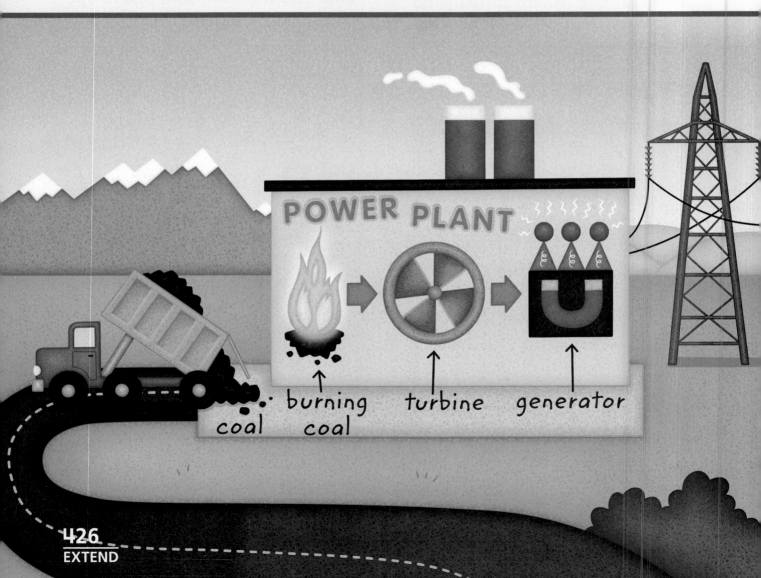

POWER PLANT

coal | burning coal | turbine | generator

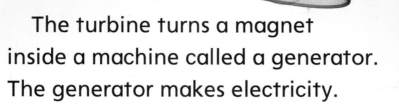

The turbine turns a magnet inside a machine called a generator. The generator makes electricity.

When you flip a switch in your home, you make a circuit with the power plant. Then, electricity flows through power lines and stations to the plug in your home and into your lamp.

Electricity leaves the power plant and travels through many power lines.

Electricity comes to my home.

I pull the cord. The light goes on.

Talk About It

Cause and Effect. What makes the light go on in your home?

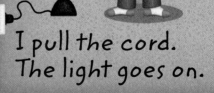

Energy Poem

Some energy can be seen.
Some energy can be felt.
Heat is a kind of energy that
can make solids melt.

Sound is a kind of energy.
I hear pitches high and low.
I hear because of vibrations,
small, big, fast, and slow.

Light is a kind of energy.
Light can bounce and bend.
I block light to make shadows.
I see white when all colors blend.

Electricity is a kind of energy.
It makes many things run.
Without electricity, my toy
would be no fun!

CHAPTER 12 Review

Vocabulary

Use each word once for items 1–5.

circuit
current electricity
reflect
static electricity
vibrate

1. Sound is made when objects ________.
2. Energy that moves through wires is called ________.
3. Energy that jumps from cloud to cloud is called ________.
4. When light bounces off objects, the light will ________.
5. This picture shows a complete ________.

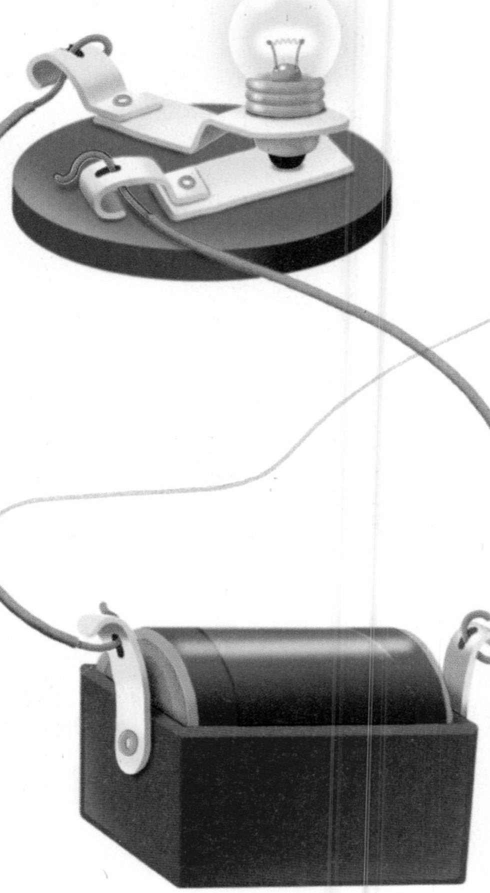

Answer the questions below.

6. What happens to a sound when it moves away from you?

7. **Measure.** How many degrees Celsius is the temperature?

8. What can heat do?

9. **Main Idea and Details.** Why can you see a rainbow with a prism?

10. How do we use energy?

Crash Tester

If you like to learn about cars and safety, you could become a crash tester. Crash testers learn how to make cars safer by setting up crashes!

These workers explore what happens to dummies, or big dolls, in a car crash. Then the crash testers decide how to make the cars safer. Crash testers study air bags and seat belts to make them protect people better.

crash tester

More Careers to Think About

mechanic

car designer

LOG ON ⊙-Careers at www.macmillanmh.com

Reference

Science Handbook

Health Handbook

Glossary

Measurements

Nonstandard

You can use objects to measure the length of some solids. Line up objects and count them. Use objects that are alike. They must be the same size.

▲ This string is about 8 paper clips long.

▲ This string is about 2 hands long.

Try It

Measure a solid in your classroom.
Tell how you did it.

Standard

You can also use a ruler to measure the length of some solids. You can measure in a unit called **centimeters**.

◄ **This toy is about 8 centimeters long. This is written as 8 cm.**

You can also use a ruler to measure in a unit called **inches**. One inch is longer than I centimeter.

◄ **This toy is about 3 inches long. This is written as 3 in.**

Try It

Estimate the length of this toy car. Then find its exact length.

Measurements

Volume

You can measure the volume of a liquid with a **measuring cup**. Volume is the amount of space a liquid takes up.

▲ This measuring cup has 1 cup of liquid.

Mass

You can measure mass with a **balance**. The side that has the object with more mass will go down.

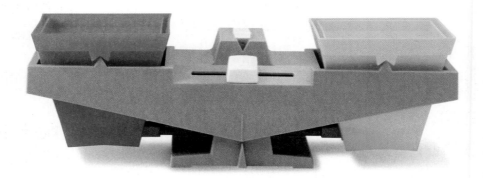

▲ Before you compare the mass of two objects, be sure the arrow points to the line.

Try It

Place two objects on a balance. Which has more mass?

Time

You can measure time with a **clock**.
A clock measures in units called hours,
minutes, and seconds. There are 60
minutes in I hour.

minute hand

hour hand

There are 5
minutes between
each number.

Temperature

Degrees
Fahrenheit

Degrees
Celsius

You can measure
temperature with
a **thermometer**.
Thermometers measure
in units called degrees.

◀ The temperature
is 85 degrees
Fahrenheit.

Try It

Use a thermometer to find the
temperature outside today.

Science Tools

Computer

A computer is a tool that can help you get information. You can use the Internet to connect to other computers around the world.

monitor

hard drive

keyboard

mouse

When you use a computer, make sure an adult knows what you are working on.

Hand Lens

A hand lens is another tool that can help you get information. A hand lens makes objects seem larger.

Try It

Use a hand lens to look at an object. Draw what you see.

Graphs

Bar Graphs

Bar graphs organize data. The title of the graph tells you what the data is about. The shaded bars tell you how much of each thing there is.

Favorite Fruits

Try It

Make a bar graph that shows your classmates' favorite fruits.

Your Body

Skeletal System

Your body has many parts. All your parts work together to help you live.

Bones are hard body parts inside your body. They help you stand straight. Your bones give your body its shape.

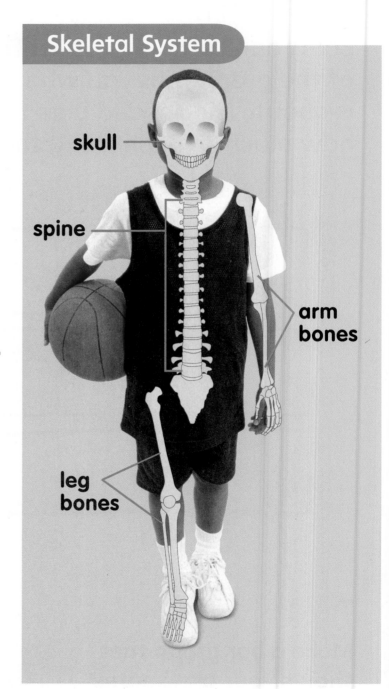

Skeletal System

skull

spine

arm bones

leg bones

Try It

How many bones do you think there are in your arms? Count them.

Muscular System

Muscles are body parts that help you move. They are inside your body.

Muscles get stronger when you exercise them.

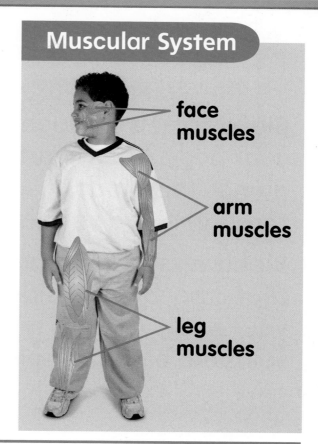

Muscular System

- face muscles
- arm muscles
- leg muscles

Nervous System

Your brain sends messages all around your body. The messages travel along tiny body parts called nerves.

These messages tell your body parts to move. They can also alert you of danger.

Try It

Jump up and down in place. Which muscles did you use?

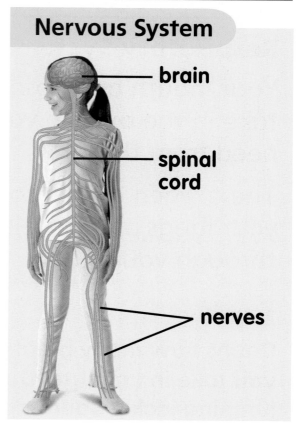

Nervous System

- brain
- spinal cord
- nerves

Your Body

Circulatory System

Blood travels through your body. Your heart pumps this blood through blood vessels.

Blood vessels are tubes that carry blood inside your body. Arteries and veins are blood vessels.

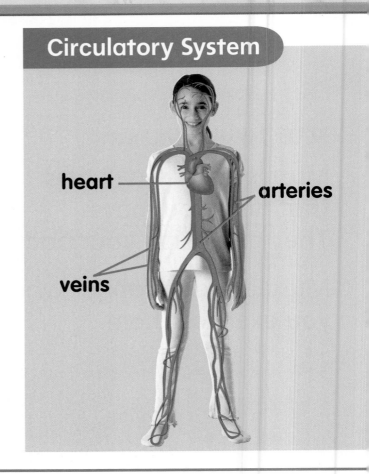

Circulatory System

heart
arteries
veins

Respiratory System

Your mouth and nose take in the oxygen you need from the air.

The oxygen goes into your lungs and travels through your blood.

Try It

Count how many breaths you take in 1 minute. Do ten jumping jacks. Count again.

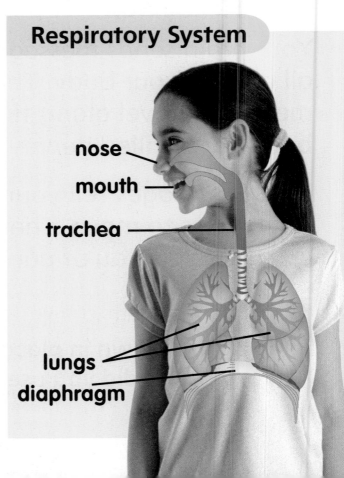

Respiratory System

nose
mouth
trachea
lungs
diaphragm

Digestive System

When you eat, your body uses food for energy. Food enters your body through your mouth. Your stomach and intestines help you get nutrients from the food in your body.

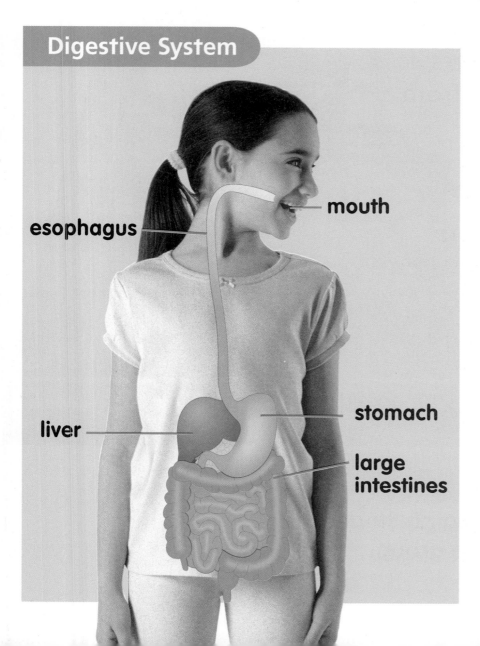

Digestive System

mouth

esophagus

liver

stomach

large intestines

Try It

Write a story about why your body needs food.

Healthful Foods

MyPyramid

MyPyramid is a guide for healthful eating. A healthful meal contains foods from the five food groups. A food group is a group of foods that are alike.

Eat more foods from the largest slice of the pyramid. Eat less from the smallest slice.

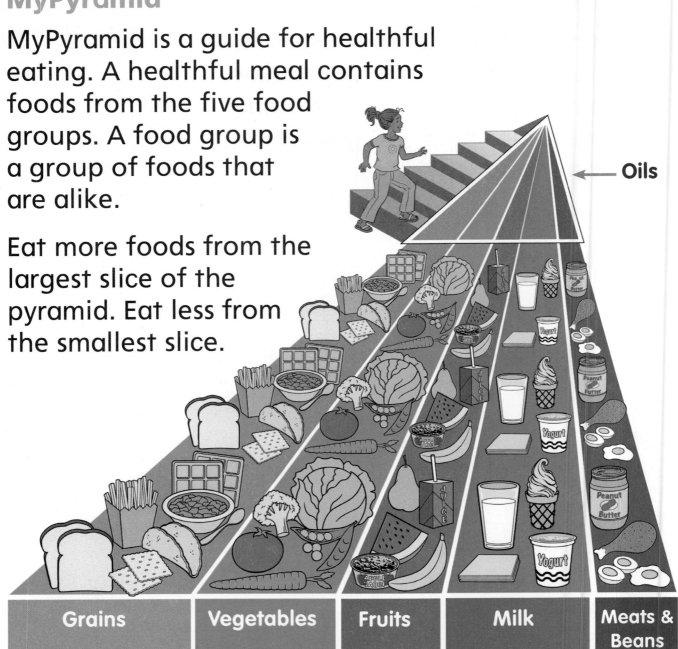

Oils

Grains | Vegetables | Fruits | Milk | Meats & Beans

Try It

Plan a healthful meal. Include one food from each group.

Healthful Foods

Nutrients are materials in foods that make you healthy. Nutrients called carbohydrates store energy in your body. Proteins help your body grow.

People around the world get nutrients from different foods.

chickpeas

Foods Around the World		
Food	Parts of the world	Nutrient
Rice	Asia	carbohydrate
Tortilla	Central America	carbohydrate
Millet	Africa	carbohydrate
chickpeas	Middle East	protein
olives	Europe	oils

rice

olives

Try It

List your favorite foods. Find out what nutrients are in the food.

Healthy Living

Stay Healthy

Be active every day. Exercise keeps your heart and lungs healthy.

Doctors and dentists can help you stay healthy as you grow.

▲ Exercise is important for a healthy body.

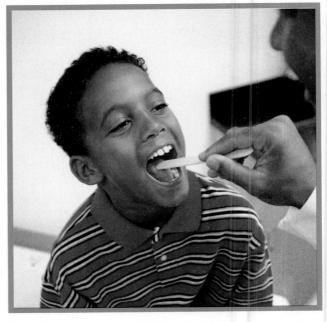

▲ Get a checkup from a doctor and dentist every year.

Try It

Record how many times you exercise in one week.

Take Care of Your Body

Tobacco and alcohol harm you. Tobacco smoke can make it hard to breathe. Alcohol slows down your mind and body.

Here are some ways to take care of your body. ▼

▲ **Only take medicines that your parent or doctor gives you.**

Take Care of Yourself

Take a bath.

Brush and floss your teeth every day.

Stand up straight.

Get plenty of sleep.

Try It

Make a poster about being drug free. Share it with your school.

Safety Indoors

To stay safe indoors, do not touch dangerous things. Tell an adult about them right away. Never taste anything without permission.

In case of a fire, get out fast. If your clothes catch fire, remember to stop, drop, and roll.

▲ **Do not touch these things.**

Try It

Practice stop, drop, and roll. Teach it to a friend.

stop

drop

roll

Safety Outdoors

Be safe outdoors. Follow these rules.

▲ Wear a helmet.

▲ Cross at a crosswalk.

▲ Wear your seat belt.

▲ Follow game rules.

Try It

Choose one of the rules. Make
a poster showing the safety rule.

Wind Scale

Very windy weather can be dangerous. Do not play outdoors before a storm. Scientists use a scale like this one to tell how hard the wind is blowing.

▼ **This is a Beaufort scale.**

Wind Scale		
Number	**What You Can See**	**Wind Speeds**
0	calm	less than 1 mile per hour
3	gentle breeze	8–12 miles per hour
6	strong breeze	25–30 miles per hour
9	very strong wind	47–54 miles per hour
12	hurricane	more than 73 miles per hour

Try It

Look out a window. What kind of wind do you see? Use the wind scale to help.

Glossary

A

adaptation Body part or a way an animal acts that helps it stay alive. (page 70) **The anteater's long snout is an adaptation.**

amphibian Animal that lives part of its life in water and part on land. (page 57) **A salamander is an amphibian.**

anemometer A tool that measures the speed of wind. (page 227) **The stronger the wind is, the faster the anemometer spins.**

Arctic A very cold place near the North Pole. (page 132) **Animals in the Arctic have layers of fat to keep them warm.**

attract To pull towards something. (page 386) **A magnet can attract some objects.**

axis A center line that an object spins around. (page 255) **Earth spins on its axis.**

camouflage   A way that animals blend into their surroundings. (page 71) **Animals use camouflage to stay safe.**

chemical change   When matter changes into different matter. (page 328) **Cooking an egg makes a chemical change.**

circuit   A path that electricity flows in. (page 422) **A bulb will light when connected with wires in a circuit.**

cirrus   Thin, wispy clouds high in the sky. (page 239) **The wind blows cirrus clouds into wispy streams.**

condense   To change from a gas to a liquid. (pages 233, 334) **Water vapor can condense on a cold glass.**

core   Earth's deepest and hottest layer. (page 160) **The core is thousands of miles below our feet.**

core

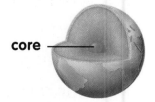

crust   Earth's outer layer. (page 160) **We live on Earth's crust.**

crust

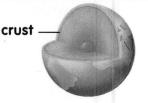

cumulus White, puffy clouds. (page 238) **Small cumulus clouds appear in good weather.**

current electricity A kind of energy that moves in a path. (page 422) **When you use a toaster, you use current electricity.**

D

decompose When plant and animal parts rot or break down. (page 198) **This log will decompose over time.**

desert A dry habitat that gets very little rain. (page 130) **A desert is hot and dry.**

dissolve To mix evenly with a liquid and form a solution. (page 343) **Sugar will dissolve when it is mixed with water.**

drought A long period of time with little or no rain. (page 104) **Plants can die in a drought.**

E

earthquake   A shake in Earth's crust. (page 174) **An earthquake damaged this road.**

endangered   When many of one kind of animal die and only a few are left. (page 106) **These tigers are endangered.**

evaporate   To change from a liquid to a gas. (pages 232, 333) **Water can evaporate from oceans, rivers, lakes, or land.**

extinct   When a living thing dies out and no more of its kind live on Earth. (page 109) **Dinosaurs are extinct.**

F

flood   Water that flows over land and cannot easily soak into the ground. (page 175) **This man is walking in a flood.**

flower   Plant part that makes seeds. (page 30) **Some flowers can grow into fruit.**

food chain A model of the order in which living things get the food they need. (page 96) **A food chain begins with the Sun.**

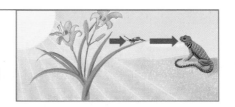

food web Two or more food chains that are connected. (page 99) **This picture shows a desert food web.**

force A push or pull on an object. (page 369) **When you kick a ball, you are using a kind of force.**

fossil What is left of a living thing from the past. (page 108) **This fish fossil was found in the desert.**

fresh water Water that is not salty. (page 165) **Fresh water is found in lakes, ponds, rivers, and streams.**

friction A force that slows down moving things. (page 371) **A skate makes friction when the wheels rub against the ground.**

fuel Material burned to make power or heat. (page 401) **Wood is fuel for fire.**

fulcrum The point that a lever moves against. (page 378) **This piece of wood can act as a fulcrum.**

fulcrum

G

gas Matter that spreads to fill the space it is in. (page 312) **The tube is filled with gas.**

gravity A kind of force that pulls down on everything on Earth. (page 370) **Gravity is the force that pulls a ball to the ground.**

H

habitat A place where plants and animals live. (page 90) **A habitat can be wet, dry, windy, or cold.**

heat Kind of energy that makes objects warmer. (page 400) **The Sun gives us heat.**

I

insect Animal with six legs, antennae, and a hard outer shell. (page 58) **An ant is an insect.**

L

landform One of the different shapes of Earth's land. (page 156) **This landform is called a valley.**

landslide Sudden movement of soil from higher to lower ground. (page 175) **Buildings can be damaged in a landslide.**

larva Stage in the life cycle of some animals after they hatch from an egg. (page 64) **A caterpillar is a larva.**

lever A simple machine made of a bar that turns around a point. (page 378) **A lever can help you move or lift objects.**

life cycle How a living thing grows, lives, has young, and dies. (pages 34, 62) **The pictures show the life cycle of a chicken.**

light A kind of energy that lets us see. (page 416) **We get light from the Sun.**

liquid Matter that takes the shape of the container it is in. (page 310) **Water is a liquid.**

M

mammal Animal with hair or fur that feeds milk to its young. (page 56) **A lion is a mammal.**

mantle Very hot layer below Earth's crust. (page 160) **The mantle is too hot for living things.**

mantle —

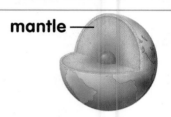

mass The amount of matter in an object. (page 296) **The larger boot has more mass.**

matter Anything that takes up space and has mass. (page 296) **Everything around us is made of matter.**

minerals Bits of rock and soil that help plants and animals grow. (pages 26, 190) **Plants use minerals in the ground to grow.**

mixture Two or more things mixed together that keep their own properties. (page 340) **This snack food is a mixture.**

motion A change in the position of an object. (page 363) **This dog is in motion.**

N

natural resource Material from Earth that people use in daily life. (page 188) **Rocks are a natural resource.**

O

ocean A large body of salt water. (pages 136, 166) **The ocean covers most of Earth.**

orbit The path Earth takes around the Sun. (page 262) **Earth orbits the Sun each year.**

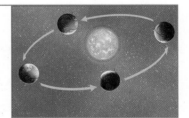

oxygen A gas found in the air we breathe. (page 27) **Living things need oxygen.**

phase The Moon's shape as we see it from Earth. (page 271) **The Moon's phase will change each night.**

physical change A change in the size or shape of matter. (page 326) **When you fold matter, you make a physical change.**

pitch How high or low a sound is. (page 409) **Short, tight strings make a high pitch.**

high pitch —
low pitch —

planet A huge object that travels around the Sun. (page 276) **Mercury is the planet closest to the Sun.**

poles The two ends of a magnet, or either end of Earth's axis. (page 388) **Earth has two poles, a north pole and a south pole.**

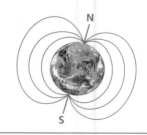

pollen Sticky powder inside a flower that helps make seeds. (page 30) **Pollen can move from flower to flower.**

pollen —

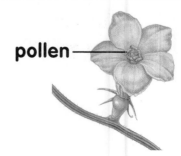

pollution   Anything that makes air, land or water dirty. (page 204) **Garbage is one kind of pollution.**

pond   A small body of fresh water. (page 138) **A pond is home to plants and animals.**

position   The place where something is. (page 362) **The position of the dog is above the cat.**

precipitation   Water falling from the sky as rain, snow, or hail. (page 225) **Rain is one kind of precipitation.**

predator   An animal that hunts other animals for food. (page 97) **A predator must be fast to catch its food.**

prey   Animals that are eaten by predators. (page 97) **The bird catches prey in its beak.**

property The look, feel, smell, sound, or taste of a thing. (page 298) **One property of this toy toucan is that it is soft.**

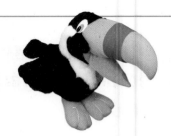

pupa Stage in a butterfly life cycle when a caterpillar makes a hard case around itself. (page 64) **The pupa hangs from a branch.**

R

rain forest A habitat where it rains almost every day. (page 124) **Many kinds of plants and animals live in a rain forest.**

ramp A simple machine with a flat, slanted surface. (page 379) **A ramp can be used to move an object from one level to another.**

recycle To make new items out of old items. (page 206) **You can recycle paper.**

reduce To cut back on how much you use something. (page 206) **We should reduce the amount of water we use.**

reflect To bounce off something. (page 416) **Light can reflect better off shiny objects.**

repel To push away or apart. (page 388) **The two south poles of a magnet repel each other.**

reptile Animal with rough, scaly skin. (page 57) **An alligator is a reptile.**

reuse To use something again. (page 206) **We can reuse items to cut down on waste.**

rock A hard, nonliving part of Earth. (page 188) **A rock like this can be used as a tool.**

rotation A turn or spin. (page 254) **Earth makes one rotation in 24 hours.**

seed Plant part that can grow into a new plant. (page 30) **A seed can grow with water, warmth, and air.**

seedling A young plant. (page 34) **A seedling will grow into an adult plant.**

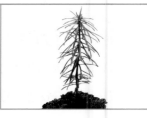

simple machine A tool that can change the size or direction of a force. (page 378) **This simple machine is called a ramp.**

soil A mix of tiny rocks and bits of dead plants and animals. (page 196) **Most plants need soil to grow.**

solar system The Sun, eight planets, and their moons. (page 276) **Planets in our solar system orbit the Sun.**

Earth

solid Matter that has a shape of its own. (page 302) **This chair is a solid.**

solution A kind of mixture with parts that do not easily separate. (page 343) **Water and this drink mix make a solution.**

sound A type of energy that is heard when objects vibrate. (page 406) **An alarm clock makes a loud sound.**

speed How far something moves in a certain amount of time. (page 364) A **cheetah has a fast running speed.**

star An object in space made of hot, glowing gases. (page 272) **The Sun is a star that we see during the day.**

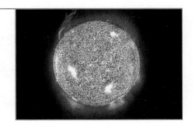

static electricity A kind of energy made by tiny pieces of matter that attract and repel each other. (page 424) **Static electricity attracts the girl's hair to the balloon.**

stratus Thin clouds that form into layers like sheets. (page 239) **Stratus clouds can cover the whole sky.**

temperature A measurement of how hot or cold something is. (page 224) **A low temperature means something is cold.**

trait The way a living thing looks or acts. (page 41) **The color of a flower is a trait.**

 V

vibrate To move back and forth quickly. (page 407) **Strings vibrate to make sound.**

volcano An opening in Earth's mantle and crust. (page 174) **A volcano can change the land quickly.**

volume The amount of space something takes up. (page 311) **You can measure the volume of a liquid with measuring cups.**

 W

woodland forest A habitat that gets enough rain and sunlight for trees to grow well. (page 122) **Many deer live in a woodland forest.**

Science Skills

classify To group things by how they are alike. (page 5) **You can classify animals by how many legs they have.**

communicate To write, draw, or tell your ideas. (page 9) **You can communicate the ways you can change a piece of clay.**

Changing Clay
1. I roll the clay.
2. I pinched the clay.
3. I squeezed the clay.
4. I poked the clay.

compare To observe how things are alike or different. (page 5) **You can compare how a cat and a dog are alike and different.**

cats   alike   dogs
meow   4 legs   bark
rough   whiskers   wet
tongue   nose

draw conclusions To use what you observe to explain what happens. (page 9) **You can draw conclusions about why the stick will make a shadow.**

infer To use what you know to figure something out. (page 7) **From these tracks, you can infer which animal was here.**

investigate To make a plan and try it out. (page 8) **You can investigate how long it takes the car to stop rolling.**

make a model To make something to show how something looks. (page 4) **You can make a model of a mountain in the ocean.**

measure To find out how far something moves, or how long, how much, or how warm something is. (page 6) **You can measure temperature with a thermometer.**

observe To see, hear, taste, touch, or smell. (page 4) **You can observe how the flower looks, smells, and feels.**

predict To use what you know to tell what you think will happen. (page 8) **You can predict what the weather will be like today.**

put things in order To tell or show what happens first, next, or last. (page 7) **You can put things in order to show the life cycle of a plant.**

record data To write down what you observe. (page 6) **You can record data about what your class had for lunch.**

Our Lunch

liquid

Solid

0 1 2 3 4 5 6 7 8 9
number of solids and liquids

Cover Photography Credits: Front Steven M. Rollman/Natural Selection Stock Photography; **Back** Tomasz Pietryszek/E+/Getty Images; **Inside Back** Tomasz Pietryszek/E+/Getty Images; **Inside Front** Tomasz Pietryszek/E+/Getty Images; **Spine** Tomasz Pietryszek/E+/Getty Images.

Photography Credits: 0-i Tomasz Pietryszek/E+/Getty Images; **iv** (tr)Ken Cavanaugh/Macmillan/McGraw-Hill, (tr-inset)George Grall/National Geographic Image Collection, (bl)Macmillan/McGraw-Hill, (br)Siede Preis/Photodisc Green/Getty Images; **ix** (l) Digital Vision/PunchStock, (c)Space Telescope Science Institute/NASA/Photo Researchers, Inc, (r)StockTrek/Getty Images; **vi** Gallo Images-Heinrich van den Berg/Photodisc/Getty Images; **vii** McGraw-Hill Companies, Inc. Mark Dierker, photographer; **viii** (bl)Ed Reschke/Getty Images, (bc)Valerie Giles/Photo Researchers, Inc, (bkgd)Nikolaj Schouboe/E+/Getty Images; **x** Ken Cavanaugh/Macmillan/McGraw-Hill; **xi** ©McGraw-Hill Education/Joe Polillio; **xii** (l) EyeWire/Getty Images, (r)Photodisc/Getty Images, (others)The McGraw-Hill Companies; **xiii** (t)Photodisc/Getty Images, (others)The McGraw-Hill Companies; **xiv** (t)Dynamic Graphics Group/IT Stock Free/Alamy, (bl)Macmillan/McGraw-Hill,(br)Comstock Images/Alamy; **xv** (t) Studio Photogram/Alamy, (cl)frenchman77, (c)Andrew J. Martinez/Photo Researchers, (bl bc) The McGraw-Hill Companies Inc./Ken Cavanagh Photographer, (br)Arco Images/Alamy; **xvi** (t) C Squared Studios/Getty Images, (b)Ken Karp/Macmillan/McGraw-Hill; **1** (bkgd)Don Paulson Photography/SuperStock; **2** (inset)Geostock/Getty Images; **2-3** (bkgd)Geostock/Getty Images; **3** (t)Photodisc/Getty Images, (b)Ken Karp/Macmillan/McGraw-Hill, (others)Macmillan/McGraw-Hill; **4** (tl)Brand X Pictures/PunchStock, (tr)Michael Orton/Getty Images, (b)Michael P. Gadomski/Photo Researchers, Inc.; **5** (l)Martin Ruegner/Getty Images, (r)©Ingram Publishing/Alamy; **6** (tl)Ken Cavanaugh/Macmillan/McGraw-Hill, (tr)U.S. Fish and Wildlife Service photo by Bonnie Strawser, (bl)Brand X Pictures/PunchStock, (br)Siede Preis/Photodisc/Getty Images, (inset)George Grall/National Geographic Image Collection; **7** (t)Siede Preis/Getty Images, (tr bl)Photodisc/Getty Images, (br)Norbert Rosing/National Geographic Image Collection; **8** (l)Jason Edwards/Media Bakery, (c)Hans Pfletschinger/Peter Arnold, Inc.; **9** Photodisc/Getty Images; **10-11** Bob Elsdale/Getty Images, (t)Design Pics/Corey Hochachka, (tc)Justus de Cuveland/Getty Images, (b)Don Farrall/Getty Images, (bc)Jeff Foott/Getty Images, (inset)Photodisc/Getty Images; **12** (l)©IT Stock/PunchStock, (r)Dan Suizo/Photo Researchers, Inc.; **12-13** Ken Karp/Macmillan/McGraw-Hill; **13** James Balog/Getty Images; **14** (l)James Balog/Getty Images, (t)Siede Preis/Photodisc/Getty Images; **15** (t)Photodisc/Getty Images, (b)©IT Stock/PunchStock, (r)Dan Suizo/Photo Researchers, Inc.; **15a** (t bl)C Squared Studios/Getty Images, (bc)Photodisc Green/Getty Images, (br)© Brand X Pictures/Alamy; **15b** Elizabeth Ballengee/MazerStock; **16** (tl)Michael Groen/Macmillan/McGraw-Hill, (others)Ken Cavanaugh/Macmillan/McGraw-Hill; **17** Ronald Wittek/Getty Images; **20-21** Frank Krahmer/Masterfile; **21** (t)Papilio/Alamy, (c)Photo Courtesy of Pdphoto.org, (b)Comstock Images/Alamy; **22** Corbis; **22-23** Gallo Images/Danita Delimont/Getty Images; **23** (t)Ken Karp/Macmillan/McGraw-Hill, (c b)Macmillan/McGraw-Hill; **24** (t)Royalty-Free/Corbis, (b) Adams Picture Library/Alamy; **25** (tl)Steve Taylor/SPL/Photo Researchers, Inc., (tc)Papilio/Alamy, (tr)C Squared Studios/Getty Images, (b)Corbis; **26** (t)Derek E. Rothchild/Getty Images, (b)Burke/Getty Images, (bkgd)Corbis; **27** ©Corbis, (b)Siede Preis/Getty Images; **28-29** Interfoto Pressebildagentur/Alamy; **29** (t)Michael Scott/Macmillan/McGraw-Hill, (others) Macmillan/McGraw-Hill; **30** Rudy Malmquist/Getty Images; **31** (t)Ingram Publishing/age fotostock, (br)Malcolm Case-Green/Alamy; **32** (l)Blickwinkel/Alamy, (cl)©Steffen Hauser/botanikfoto/Alamy, (cr)The McGraw-Hill Companies, Inc./Jacques Cornell photographer, (r) Martin Ruegner; **33** (tr)DAJ/Getty Images, (c bc)Ken Cavanaugh/Macmillan/McGraw-Hill; **34** (tl)©Steven P. Lynch/The McGraw-Hill Companies, (tc)McGraw-Hill Education; **34** (tr)Nuno Tavares, (bl)David R. Frazier Photolibrary, Inc./Alamy, (br)Comstock Images/Alamy; **35** (l)Peter Arnold Inc./Alamy, (r)Design Pics/PunchStock; **36** imagestate/Alamy; **37** (l)Pixtal/age fotostock, (c)DIOMEDIA/Alamy, (r)PhotoAlto/PunchStock; **38-39** Tim Laman/Getty Images; **39** (l)Ken Karp/Macmillan/McGraw-Hill, (others)Macmillan/McGraw-Hill; **40** (tr)Brand X Pictures/PunchStock, (cl)Siede Preis/Getty Images, (cr)Papilio/Alamy, (b)Gary Crabbe/Alamy; **41** (t)Brand X Pictures/PunchStock, (b)Russell Illig/Getty Images; **42** (l)David Noton/Getty Images, (r)Stocktrek; **42-43** Fabrice Bettex/Alamy; **45** (l)Edward Parker/Alamy, (r)Dominique Halleux/Peter Arnold, Inc.; **46** (inset)Peter Fakler/Alamy, (bkgd)Arco Images/Alamy; **47** Wilson Goodrich/Index Stock Imagery; **48** (inset)Thomas R. Fletcher/Getty Images, (bkgd) Masterfile; **49** (inset)©Alfred Schauhuber/age fotostock, (bkgd)Danita Delimont/Alamy; **50** (l)Macmillan/McGraw-Hill, (r)Photo Courtesy of Pdphoto.org; **51** (tl)Jim Allan/Alamy, (tr)John Henwood/Alamy, (b)Don Paulson Photography/Purestock/SuperStock; **51A** Macmillan/McGraw-Hill; **52-53** Gail Shotlander/Getty Images; **53** (t)franzfoto.com/Alamy, (tcr)©Central Stock/Fotosearch, (tcl)Jane Burton/Dorling KindersleyGetty Images, (tc)©Photodisc/Getty Images, (bc)U.S. Fish & Wildlife Service, (b)McGraw-Hill Education; **54-55** PHOTO BY PRASIT CHANSAREEKORN/Moment Open/Getty Images; **55** (t)National Geographic/Getty Images, (tcl)Alan and Sandy Carey/Getty Images, (tcr)Richard F. Wintle/Getty Images, (b)Julie Bedford, NOAA PA, (bcl)© Helder Joaquim Soares Almeida/SuperFusion/SuperStock, (bc)Andrew M. Snyder/Getty Images; **56** franzfoto.com/Alamy; **57** (tl)Herve Berthoule/Jacana/Photo Researchers, Inc., (tr)Danita Delimont/Alamy, (cr)Bruce Coleman, Inc./Alamy, (b)Clark Wheeler; **58** (l)Kick Images/Getty Images, (r)Corbis; **59** (t)Bob Elsdale/Getty Images, (c)Exactostock/SuperStock, (b)Tim Ridley/Getty Images; **60-61** Mark Malkinson/Alamy; **61** Nancy Ney/Digital Vision/Getty Images; **62** (tc)Hanquan Chen/Getty Images, (tr)Raymond Liu , Hong Kong/Moment/Getty Images, (br)©Central Stock/Fotosearch, (bl)Gina Guarnieri/E+/Getty Images, (bc)©Photodisc/Getty Images, (bc)©Feng Wei Photography, (b)©Siede Preis/Getty Images; **64** (l)Papilio/Alamy, (c)U.S. Fish & Wildlife Service, (r)John T. Fowler/Alamy; **65** (l)Phototake/Alamy, (r)DIOMEDIA/Alamy; **66** Courtesy of Nancy Simmons; **66-67** (bkgd)Sundell Larsen/Photodisc/Getty Images; **67** (t)©Alex Joukowski/Getty Images, (b)Wildlife Pictures/Peter Arnold, Inc.; **68-69** Gallo Images-Heinrich van den Berg/Photodisc/Getty Images; **69** (t)Ken Karp/Macmillan/McGraw-Hill, (others) Macmillan/McGraw-Hill; **70** (l)Comstock/PunchStock, (tr)Magdalena Biskup Travel Photography/Getty Images, (b)Dynamic Graphics Group/IT Stock Free/Alamy, (br)McGraw-Hill Education; **71** (tl)Design Pics/Natural Selection Robert Cable, (tc)Design Pics/Corey Hochachka, (tr)©Chase Swift/Corbis, (b)Royalty-Free/Corbis; **72** (t)Bob Elsdale, (tr)Fuse, (b)Georgette Douwma/Getty Images; **73** (tl)Hemera/age fotostock, (tr)Creatas/PunchStock, (b)PhotoDisc/Getty Images; **74** (l)©Stockbyte/PunchStock, (r)Radius Images/Alamy; **75** Jupiterimages; **76-77** ©imagebroker.net/SuperStock; **77** Armin Floreth/imagebroker/Alamy; **78** Graeme Purdy/Getty Images; **78-79** Ingram Publishing/SuperStock; **80** (t)U.S. Fish & Wildlife Service, (b)Royalty-Free/Corbis; **81** (tl)Liam Douglas, (tr)Corbis, (cr)Photodisc/Getty Images, (bcl) DIOMEDIA/Alamy, (bcr)Papilio/Alamy, (br)John T. Fowler/Alamy; **81a** (tl)Corbis, (tr)©Stockbyte/PunchStock, (bl)©McGraw-Hill Companies, Inc./Carlyn Iverson, photographer, (br)Dana Hursey/Masterfile; **82** (t)John and Karen Hollingsworth/U.S. Fish and Wildlife Service, (bl)Chuck Place/Alamy, (br)Cleveland Metroparks Zoo/McGraw-Hill Education; **83** Ted Mead/Getty Images; **86-87** ©Charles Mauzy/Corbis; **87** (t)McGraw-Hill Companies, Inc. Mark Dierker, photographer, (tc)Brand X/Corbis, (bc)Thinkstock/Jupiterimages, (b)©McGraw-Hill Companies, Inc./Carlyn Iverson, photographer; **88-89** Georgette Douwma/Getty Images; **89** (t c)Macmillan/McGraw-Hill, (b)Spike Mafford/Getty Images; **90** (t)McGraw-Hill Companies, Inc. Mark Dierker, photographer, (b)Tim Laman/Getty Images; **90-91** Klein/Peter Arnold, Inc.; **91** ESO/H.H.Heyer, CC BY 3.0; **93** The McGraw-Hill Companies, Inc./Janette Beckman photographer; **94-95** Brand X/Corbis; **95** (t c)Macmillan/McGraw-Hill, (bl)Royalty-Free/Corbis, (bc)Image Source/Getty Images, (br)David De Lossy/Getty Images; **99** ©Stockbyte; **101** (tl)George Grall/Getty Images, (tr cl b)Creatas/Punchstock, (c cr)Dynamic Graphics Group/IT Stock Free/Alamy; **102-103** Stacey Newman/Getty Images; **103** (br)Ken Karp/Macmillan/McGraw-Hill, (others)Macmillan/McGraw-Hill; **104** Thinkstock/Jupiterimages; **104-105** (bcr)BRUCE COLEMAN INC./Alamy; **105** (inset)Catherine Ledner/Getty Images, (t)Jim Zuckerman/Corbis; **106** Schafer & Hill/Getty Images; **106-107** Nancy Nehring/Photodisc/Getty Images; **107** Steven Trainoff Ph.D./Getty Images; **108** (inset)©McGraw-Hill Companies, Inc./Carlyn Iverson, photographer, (b)James L. Amos/National Geographic Image Collection; **109** Alan Morgan; **110** Dennis Finnin/American Museum of Natural History; **110-111** (bkgd)Corbis; **111** Courtesy American Museum of Natural History; **112-113** C Squared Studios/Getty Images; **113** (t)Corbis, (b)Photodisc/Getty Images; **114-115** C Squared Studios/Getty Images; **115** PhotoDisc/Getty Images; **116** ©Stockbyte; **117** (l)Corbis, (r)ThinkStock/Jupiterimages; **118-119** Design Pics/Robert Brown; **119** (tc)Steven Kazlowski/Peter Arnold, Inc., (b)Tim Krieger/Alamy, (bc)Ingram Publishing/age fotostock; **120-121** Perry Mastrovito/Jupiterimages; **121** (l)Ken Karp/Macmillan/McGraw-Hill, (others)Macmillan/McGraw-Hill; **122** (inset)©Ronald Wittek/Getty Images; **122-123** Photo by Jeff Vanuga, USDA Natural Resources Conservation Service; **123** (l to r, t to b) Photodisc/Getty Images, NPS Photo by Jim Peaco, Pixtal/AGE Fotostock, Getty Images, Punchstock, © SuperStock/Alamy, ©Siede Preis/Photodisc/Getty Images; **124** (t)Dynamic Graphics Group/IT Stock Free/Alamy, (c)Lee Feldstein/Alamy, (b)©Comstock/PunchStock; **124-125** Niall Corbet @ www.flickr/photos/niallcorbet/Getty Images; **125** (t)©McGraw-Hill Companies, Inc./Carlyn Iverson, photographer, (b)Design Pics/Natural Selection Ralph Curtin; **126** (tr)Courtesy of Rebecka Eriksson/American Museum of Natural History, (b)© National Geographic Image Collection/Alamy; **126-127** Ron Giling/Peter Arnold, Inc.; **128-129** Jack Goldfarb/DesignPics/Getty Images; **129** (t)Ken Karp/Macmillan/McGraw-Hill, (others) Macmillan/McGraw-Hill; **131** (tl)Steve Cooper/Photo Researchers, Inc., (tr)Design Pics/SuperStock, (bl)Derrick Hamrick/Getty Images, (br)Design Pics/David Ponton; **132** (t)Steven Kazlowski/Peter Arnold, Inc., (b)Phil A. Dotson/Photo Researchers, Inc.; **132-133** ©Ingram Publishing/Alamy; **133** ajliikala/Getty Images; **134-135** Ingram Publishing/age fotostock; **135** (l)Ken Karp/Macmillan/McGraw-Hill, (others)Macmillan/McGraw-Hill; **137** (tl)©Don Hammond/Design Pics/Corbis, (tr)Stephen Frink/Getty Images, (c)Kelvin Aitken/Peter Arnold,